T0275828

LONDON MATHEMATICAL SOCIETY LECTURE NOTE SERIES

Managing Editor: Professor J.W.S. Cassels, Department of Pure Mathematics and Mathematical Statistics, University of Cambridge, 16 Mill Lane, Cambridge CB2 1SB, England

The books in the series listed below are available from booksellers, or, in case of difficulty, from Cambridge University Press.

LONDON MATHEMATICAL SOCIETY LECTURE NOTE SERIES

Managing Editor: Professor J.W.S. Cassels, Department of Pure Mathematics and Mathematical Statistics, University of Cambridge, 16 Mill Lane, Cambridge CB2 1SB, England

The books in the series listed below are available from booksellers, or, in case of difficulty, from Cambridge University Press.

London Mathematical Society Lecture Note Series. 163

Topics in Varieties of Group Representations

Samuel M. Vovsi
Rutgers University

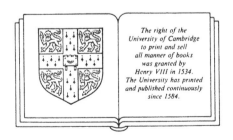

The right of the
University of Cambridge
to print and sell
all manner of books
was granted by
Henry VIII in 1534.
The University has printed
and published continuously
since 1584.

CAMBRIDGE UNIVERSITY PRESS

Cambridge

New York   Port Chester   Melbourne   Sydney

Published by the Press Syndicate of the University of Cambridge
The Pitt Building, Trumpington Street, Cambridge CB2 1RP
40 West 20th Street, New York, NY 10011-4211, USA
10 Stamford Road, Oakleigh, Victoria 3166, Australia

First published 1991

British Library cataloguing in publication data available

Library of Congress cataloguing in publication data available

ISBN 0 521 42410 0

Transferred to digital printing 2004

To my father

CONTENTS

PREFACE

The study of varieties of algebraic structures, i.e. classes of algebraic structures definable by identical relations, was originated by G. Birkhoff [7] and B. H. Neumann [67] in the 1930's. A wide expansion of the ideas and methods of variety theory began in the 1950's, when the work of G. Higman, A. I. Mal'cev, B. H. Neumann, H. Neumann, A. Tarski, W. Specht was particularly influential. Since then the intensity of work in this area of algebra has remained very high, and the number of publications devoted to identities and varieties of algebraic structures is now counted in the thousands. Various aspects of the field have been systematically presented in numerous monographs and surveys — see for example [3, 12, 15, 16, 18, 64, 68, 80, 83, 84, 85, 86, 105].

As a result of this expansion, at the present moment one can speak of a number of independent but closely related algebraic theories: varieties of groups, varieties of associative algebras and polynomial identities, varieties of Lie algebras, varieties of semigroups, varieties of lattices and universal algebras, and others. We say "independent" because each of these fields has its own motivations and stimuli for development and its own natural problems. On the other hand, they are developing in close interconnection and are constantly influencing one another.

The present book is devoted to one of the newest branches of variety theory: varieties of group representations. This subject has existed for about twenty years; its foundations were laid in papers of B. I. Plotkin and his students in the late 1960's–early 1970's. There are many motivations for the study of varieties of group representations. First, from the standpoint of universal algebra all representations of groups over a given commutative ring form a variety of two-sorted algebras, with quite natural free objects, verbal subobjects and other standard attributes of variety theory. Second, a number of classical theorems and problems are, in fact, concerned with group representations satisfying certain identical relations and, as a result, can be naturally interpreted in the framework of the theory of varieties. Here one can mention a theorem of Kolchin [42] on triangulability of a unipotent matrix group, a theorem of Kaloujnine [38] on nilpotency of a stable group of automorphisms, and a series of problems and results concerning the augmentation ideal and dimension subgroups. Third, the theory of varieties of group representations has numerous connections with varieties of groups, varieties of associative and Lie algebras, group rings, etc., which can provide important applications in both directions. This was first demonstrated by Bryce [9] and Ol'shansky [71] who applied the technique of varieties of group representations to investigating varieties of abstract groups.

However, in contrast with "usual" algebras — groups, rings, etc. — representations of groups are *two-sorted* algebraic structures, i.e. they have two underlying sets. Therefore the theory of varieties of group representations exibits a number of characteristic features and essentially new problems. During the last decades the importance of many-sorted algebraic structures has been constantly increasing, largely because of numerous applications in the theory of automata, data banks, theoretical computer science, etc. Group representations are, probably, the most comprehensively studied objects of this type, and certain methods of the theory of varieties of group representations can be (and have been) successfully applied to the study of other closely related objects, such as representations of semigroups, representations of associative and Lie algebras, group actions on rings, and linear automata. Among these related fields, one should especially mention the rapidly expanding theory of varieties of representations of Lie algebras, presented in the recent book of Razmyslov [84].

The theory of varieties of group representations has been developing steadily since its inception, mainly in the USSR. During the first ten years about eighty papers were published in the field, including two important surveys of Plotkin [76, 77]. The results of this period were, to a certain extent, summarized in a monograph of Plotkin and the author [80], published in 1983 and until now the only detailed exposition of the subject.

This monograph was mostly devoted to the general framework of the theory which by the beginning of the 1980's was well developed and stable. At the same time, a number of more specialized and advanced topics were not covered in [80].

In the present notes several of these topics are considered. In no way does this book pretend to be a comprehensive treatment of the subject (as is clear enough from its title). Our aim is to familiarize the reader with the current state of knowledge in the areas we treat, to establish a number of interesting results, and to attract attention to several promising problems. The material presented here is quite recent: the greater part of the results have appeared within the last four years. Of course, the choice of material reflects the personal taste of the author.

The book starts with a preparatory Chapter 0 in which we collect all the necessary definitions, notation and facts from the theory of varieties of group representations. As a result, one can read this text independently of [80]. The reader familiar with the foundations of the theory can skip this chapter without any harm.

The main body of the book consists of four chapters. Chapter 1 deals with stable varieties. These varieties play the role analogous to that of nilpotent varieties of groups and linear algebras, and are among the most investigated in the field. A large part of the chapter is concerned with the important property of homogeneity, which goes back to Mal'cev [61]. In particular, one of the principal results states that over a field of characteristic zero there exists a canonical one-to-one correspondence between all varieties of associative algebras and the so-called homogeneous Magnus varieties of group representations, under which n-nilpotent varieties of algebras and n-stable (homogeneous) varieties of representations correspond to each other. This correspondence makes it possible to involve the well developed theory of varieties of algebras in the investigation of varieties of group representations. The main technical tool is the embedding of a free group into the algebra of formal power series, discovered by Magnus [57]. Somewhat isolated in the first chapter is § 1.6, where the main role belongs to connections with Lie algebras. Using Zel'manov's theorem [105] on the nilpotency of Lie algebras with the Engel identity, we show there that over a field of characteristic zero every unipotent variety is stable. In other words, the identity $(x - 1)^n$ implies the identity $(x_1 - 1)(x_2 - 1)\ldots(x_N - 1)$ for some $N = N(n)$.

Chapter 2 is concerned with locally finite and locally finite-dimensional varieties. Such varieties are generated by their finite and, respectively, finite-dimensional representations, and this fact results in the character of the techniques used: periodic matrix groups, critical representations, irreducible representations of finite groups, etc. Generally speaking, the questions discussed here are rather traditional for variety theory, and the material of the chapter was developed under the strong influence of the theory of group varieties, in particular, of the excellent book of Hanna Neumann [68]. Among the results of Chapter 2, one can mention a nice and somewhat unexpected characterization of locally finite and locally finite-dimensional varieties, presented in § 2.2.

Chapter 3 is entirely devoted to the finite basis problem. The varieties considered there are locally finite or locally finite-dimensional, so that this chapter is a natural continuation of the previous one. Its main result is a theorem of the author and Nguyen Hung Shon [101] asserting that every stable-by-finite representation has a finite basis for its identities. It implies, in particular, that every representation of a finite group is finitely based. The proof uses to the full extent the machinery of critical objects and Cross varieties, as developed in papers of Oates–Powell [69] and Kovács–Newman [44], and a number of other technical tools specific to group representations.

Our final Chapter 4 does not focus on a single topic. Rather, it provides a selection of mutually independent results, including the most recent, which are of interest in their own right. Sections 4.1–4.2 are devoted to the finite basis problem for varieties satisfying multilinear identities, and to some natural connections between multilinear identities of group representations and those of associative algebras. In particular, using these connections and Kemer's solution of the Specht Problem [40], we deduce the following, *a priori* far from evident, fact: every system of multilinear identities of group representations over a field of characteristic zero is finitely based. Several other results show that if a variety satisfies a certain multilinear identity, then it is finitely based. Sections 4.3–4.4 deal with pure varieties over integral domains. Among the results of these sections, we mention an interesting theorem of G. M. Bergman on the product of ideals in free group rings which, in particular, gives an alternative proof of an earlier result of the author [94]: the product of pure varieties over a Dedekind domain is pure. An example outlined at the end of § 4.3 shows that this result cannot be generalized to arbitrary integral domains. Finally, the purpose of § 4.5 is to demonstrate that the theory of varieties of group representations has feedback to other fields of algebra. In this section we present a brief overview of some applications of our theory to varieties of groups, varieties of rings, and dimension subgroups.

For more detailed information concerning the contents of the book, the reader is referred to the introductory remarks at the beginning of each chapter.

One of the appealing features of the theory of varieties of group representations is the abundance and diversity of open problems. Some of these problems are mentioned in the present work, while many other can be found in [80]. Another attraction of the field is the broad range of techniques used. Here one can meet methods of traditional representation theory and the Fox free differential calculus, polynomial identities of rings and critical representations, the Magnus theory of the free group and connections with Lie algebras, free ideal rings and the classical Burnside theorems on matrix groups. Therefore we hope that the field treated in this book will be of interest to specialists in various branches of algebra.

We have tried to make this book accessible to a broad spectrum of readers including graduate students. A good graduate course in algebra (at the level of, say, Jacobson's *Basic Algebra*) plus some knowledge of variety theory should provide the necessary background. The exposition is, as a rule, detailed and complete, although, as in any more or less advanced course, it was impossible to make it absolutely self-contained. Sometimes we use without proofs certain facts from the theory of groups, rings, modules, etc., but in such cases precise references to the literature are always provided.

Despite its modest size, this book has had a difficult history. I worked on it during a very strained period in my life, and it is unlikely that I would have overcome the numerous obstacles without the help of many friends and colleagues. In the first place I am deeply indebted to Gregory Cherlin, Richard Lyons, Matthew Miller, Simon Thomas and Robert Wilson, who carefully read various portions of the manuscript and made numerous corrections and improvements: grammatical, stylistic and sometimes even mathematical. I am grateful to George Bergman for valuable and stimulating conversations in connection with § 4.3; to David Rohrlich and Eugene Speer for their help during my struggles with English grammar and various versions of TEX; and to the referee for an interesting and useful report. Finally, I would like to express my appreciation to Arkadiĭ Slin'ko and especially to Earl Taft, who helped me to recover the first draft of the manuscript after I had been separated from it for a long time.

S. M. VOVSI

New Brunswick, New Jersey
July 1991

Chapter 0

PRELIMINARIES

This introductory chapter provides a brief survey of basic concepts and facts from the theory of varieties of group representations which will be necessary for reading the main body of the book. The presentation is concise but, in general, self-contained: only a few standard and routine proofs are omitted. For a more detailed exposition of the foundations of the theory the interested reader is referred to Chapter 1 of [80].

0.1. Representations

Let K be an arbitrary but fixed commutative ring with unit which will usually be referred to as the *ground ring*. Consider a linear representation of a group G on a (left unitary) K-module V, that is, a group homomorphism $\rho : G \to \operatorname{Aut}_K V$. We suppose that elements of $\operatorname{Aut}_K V$ act on V on the right; then, denoting for any $v \in V$, $g \in G$

$$v \circ g = v\,\rho(g),$$

we obtain an *action* of G on V, i.e. a map $(v,g) \mapsto v \circ g$ from $V \times G$ to V satisfying the following conditions:

(i) for every $g \in G$ the map $v \mapsto v \circ g$ is an automorphism of V;

(ii) $(v \circ g_1) \circ g_2 = v \circ (g_1 g_2)$ for every $v \in V$, $g_1, g_2 \in G$.

Conversely, suppose there is given an action $\circ$ of a group G on a K-module V. Denoting for every $g \in G$ the map $v \mapsto v \circ g$ by $\rho(g)$, we evidently obtain a representation $\rho : G \to \operatorname{Aut} V$.

Thus from the standpoint of *many-sorted* universal algebra, a representation of a group over K can be treated as a two-sorted algebraic structure $(V, G, \circ)$, where V is a K-module, G a group, and $\circ$ an action of G on V. We will not adopt this standpoint in the present book, although it might be useful to keep it in mind, at least with respect to possible analogies, perspectives and generalizations.

Throughout these notes, a representation $\rho : G \to \operatorname{Aut} V$ is usually denoted by $\rho = (V, G)$; V is the *module* and G is the *acting group* of ρ. If $\rho = (V, G)$ is a representation, then its *kernel* $\operatorname{Ker} \rho = \operatorname{Ker}(V, G)$ is the kernel of the corresponding homomorphism $\rho : G \to \operatorname{Aut} V$; ρ is called *faithful* if $\operatorname{Ker} \rho = \{1\}$. In this case G can be considered as a subgroup of $\operatorname{Aut} V$. A representation $\rho = (V, G)$ is called *trivial*, or a *unit* representation, if $\operatorname{Ker} \rho = G$. If $V = \{0\}$, then ρ is said to be a *zero* representation.

In general, let $\rho = (V, G)$ be an arbitrary representation and $H = \operatorname{Ker} \rho$. Then the group $G/H = \bar{G}$ acts on V by the rule $v \circ gH = v \circ g$. We obtain a representation $\bar{\rho} = (V, \bar{G})$ which is certainly faithful; it is called the *faithful image* of ρ.

The group algebra of G over K is denoted by KG. The group G acts on KG by right multiplication giving the (right) *regular representation* (KG, G) which is denoted by $\operatorname{Reg}_K G$. If $\rho = (V, G)$ is any representation, then the action of G on V induces the action of KG on V by the natural rule: for arbitrary $v \in V$ and $u = \sum \lambda_i g_i \in KG$

$$v \circ u = \sum \lambda_i (v \circ g_i).$$

Therefore V can be regarded as a right KG-module (when it is clear from the context, we often say "G-module" instead of "KG-module"). Conversely, a KG-module structure on V determines an action of G on V, i.e. a representation (V, G).

Let $\rho = (V, G)$ and $\sigma = (W, H)$ be arbitrary representations over K. A *homomorphism* $\mu : \rho \to \sigma$ is a pair consisting of a homomorphism of K-modules $\mu : V \to W$ and a homomorphism of groups $\mu : G \to H$ (it is convenient to denote both these maps by a single letter) such that

$$\forall v \in V, g \in G : \quad (v \circ g)^\mu = v^\mu \circ g^\mu. \tag{1}$$

The class of all group representations over K together with all homomorphisms forms a category denoted by REP-K. It is easy to verify that a homomorphism $\mu : (V, G) \to (W, H)$ is a monomorphism (epimorphism, isomorphism) in REP-K if and only if both $\mu : V \to W$ and $\mu : G \to H$ are monomorphisms (epimorphisms, isomorphisms). If $\rho = (V, G)$ is a representation, H a subgroup of G and W an H-submodule of V, then there naturally arises a representation (W, H) called a *subrepresentation* of ρ. Clearly σ is a subrepresentation of ρ if and only if there exists a monomorphism $\sigma \to \rho$.

Let $\mu : (V, G) \to (W, H)$ be a homomorphism and let $V_0 = \mathrm{Ker}(V \to W)$, $G_0 = \mathrm{Ker}(G \to H)$. It is easy to see that (V_0, G_0) is subrepresentation of (V, G), and that the following conditions are satisfied:

(i) $G_0 \triangleleft G$;

(ii) V_0 is a G_0-submodule of V;

(iii) the induced action of G_0 on V/V_0 is trivial.

The subrepresentation (V_0, G_0) is said to be the *kernel* of the homomorphism μ and is denoted by $\mathrm{Ker}\,\mu$.

On the other hand, let $\rho_0 = (V_0, G_0)$ be a subrepresentation of $\rho = (V, G)$ satisfying (i)–(iii). Then the group G/G_0 acts on the module V/V_0 by the rule

$$(v + V_0) \circ (gG_0) = v \circ g + V_0,$$

and we obtain a *factor-representation* $\rho/\rho_0 = (V/V_0, G/G_0)$. There exists a canonical epimorphism $\kappa : \rho \to \rho/\rho_0$ whose kernel is ρ_0, and usual arguments show that every epimorphic image of ρ can be realized in such a way.

A homomorphism $\mu : (V, G) \to (W, H)$ is called *right* if $\mu : V \to W$ is an isomorphism. Up to isomorphism, we may assume that a right homomorphism acts identically on the left side of the representation. For example, the canonical epimorphism of a representation $\rho = (V, G)$ on its faithful image $\bar{\rho} = (V, G/\mathrm{Ker}\,\rho)$ is a right epimorphism. Furthermore, it is clear that every right epimorphic image of ρ is isomorphic to some factor-representation $(V, G/H)$ where $H \subseteq \mathrm{Ker}\,\rho$. Hence the faithful image of any representation is its "smallest" right epimorphic image.

Two representations are said to be *equivalent* if their faithful images are isomorphic. The fact that representations ρ and σ are isomorphic or equivalent is denoted by $\rho \simeq \sigma$ or $\rho \sim \sigma$ respectively. In this notation

$$\rho \sim \sigma \iff \bar{\rho} \simeq \bar{\sigma}.$$

Note. In the classical theory of group representations, two representations (V, G) and (W, G) of a *fixed* group G are called equivalent if there exists an isomorphism $\mu : V \to W$ such that $(v \circ g)^\mu = v^\mu \circ g$ for all $v \in V, g \in G$ (cf. (1)) . In the category REP-K of representations of *arbitrary* groups this notion becomes a particular case of isomorphism, and is not very useful. The definition of equivalent representations adopted in these notes is motivated by the following observation: any two representations with isomorphic faithful images originate from the same faithful representation, i.e. *from the same action*. Therefore, as far as one is concerned with abstract properties of group actions, two representations with isomorphic faithful images should be treated as "equivalent" in some natural sense.

Let $\rho_i = (V_i, G_i)$, $i \in I$, be arbitrary representations. Denote by $V = \overline{\prod} V_i$ the Cartesian product of the K-modules V_i and by $G = \overline{\prod} G_i$ the Cartesian product of the groups G_i. Then G acts on V componentwise and so there arises a representation (V, G) which is called the *Cartesian product* of the representations ρ_i and is denoted by $\overline{\prod} \rho_i$. It is easy to see that ρ is the product of the objects ρ_i in the category REP-K. On the other hand, if we take the (restricted!) direct sum of modules $\bigoplus V_i$ and the direct product of groups $\prod G_i$, then the naturally arising representation $(\bigoplus V_i, \prod G_i)$ is called the *direct product* of the ρ_i's and is denoted by $\prod \rho_i$.

An *operation* on classes of group representations is a function U assigning to every class $\mathcal{X}$ of representations a class U$\mathcal{X}$ such that

$$\mathcal{X} \subseteq \mathrm{U}\mathcal{X} \subseteq \mathrm{U}\mathcal{Y}$$

whenever $\mathcal{X} \subseteq \mathcal{Y}$. The *product* of operations is defined by the natural rule $(\mathrm{UV})\mathcal{X} = \mathrm{U}(\mathrm{V}\mathcal{X})$. An operation U is called a *closure operation* if $\mathrm{U}^2 = \mathrm{U}$; in this case the class U$\mathcal{X}$ is U-closed for every $\mathcal{X}$, that is $\mathrm{U}(\mathrm{U}\mathcal{X}) = \mathrm{U}\mathcal{X}$.

From now on we fix the notation of several closure operations. Namely, if $\mathcal{X}$ is an arbitrary class of representations, then:

$S\mathcal{X}$ is the class of all subrepresentations of $\mathcal{X}$-representations (i.e. of representations from $\mathcal{X}$);

$Q\mathcal{X}$ is the class of all homomorphic images of $\mathcal{X}$-representations;

$C\mathcal{X}$ is the class of Cartesian products of $\mathcal{X}$-representations;

$D\mathcal{X}$ ($D_0\mathcal{X}$) is the class of direct products (of a finite number) of $\mathcal{X}$-representations;

$V\mathcal{X}$ is the class of all representations ρ such that there exists a right epimorphic image of ρ belonging to $\mathcal{X}$.

0.2. Identities and varieties

THE MAIN CONCEPTS AND EXAMPLES. Let F be the absolutely free group of countable rank with free generators $x_1, x_2, \ldots$, KF its group algebra over the ground ring K, and $u(x_1, \ldots, x_n) = \sum \lambda_i f_i(x_1, \ldots, x_n)$ an element of KF. Suppose there is given a representation $\rho = (V, G)$ over K. If $g_1, \ldots, g_n \in G$ then $u(g_1, \ldots, g_n)$, being an element of the group algebra KG, acts naturally on V. We say that the formula

$$y \circ u(x_1, \ldots, x_n) \equiv 0$$

is an *identity* of the representation ρ if

$$v \circ u(g_1, \ldots, g_n) = 0$$

for any $v \in V$ and $g_1, \ldots, g_n \in G$. For brevity, the element $u = u(x_1, \ldots, x_n)$ is also said to be an identity of ρ. In other words, $u \in KF$ is an identity of $\rho = (V, G)$ if, for arbitrary $g_1, \ldots, g_n \in G$,

$$u(\rho(g_1), \ldots, \rho(g_n)) = 0$$

in $\text{End}_K V$. If $\mathcal{X}$ is a class of representations, then $u \in KF$ is called an identity of $\mathcal{X}$ if it is an identity of every representation from $\mathcal{X}$.

0.2.1. Definition. *A class of group representations is called a* variety *if it consists of all representations satisfying a certain set of identities.*

A two-sided ideal of the group algebra KF is said to be *fully invariant* (or *completely invariant*, or *verbal*), if it is invariant under all endomorphisms of KF induced by endomorphisms of the group F. Note that a fully invariant ideal need not be invariant under all endomorphisms of the K-algebra KF. For example, let Δ be the *augmentation ideal* of KF, that is, the ideal generated by all elements $f - 1$, $f \in F$. Clearly it is fully invariant. On the other hand, Δ is not invariant under any endomorphism of KF taking a free generator x_i to an invertible element $\lambda \neq 1$ of K.

The significance of fully invariant ideals for our theory is illustrated by the following theorem. For every class of representations $\mathcal{X}$ denote by $\mathcal{X}^\alpha$ the set of all its identities in KF. Conversely, for every subset U of KF denote by U^β the class of all representations satisfying the identities from this subset.

0.2.2. Theorem. *The maps α and β determine a Galois correspondence between classes of group representations and subsets of KF. The closed elements under this correspondence are precisely the varieties of group representations over K and the fully invariant ideals of KF.*

P r o o f. It is evident that the maps α and β satisfy the following conditions:

(i) $\mathcal{X}_1 \subseteq \mathcal{X}_2 \Longrightarrow \mathcal{X}_1^\alpha \supseteq \mathcal{X}_2^\alpha$, $\quad U_1 \subseteq U_2 \Longrightarrow U_1^\beta \supseteq U_2^\beta$;

(ii) $\mathcal{X}^{\alpha\beta} \supseteq \mathcal{X}$, $\quad U^{\beta\alpha} \supseteq U$.

This exactly means that the maps α and β determine a Galois correspondence (see for example [12, Ch.2, §1]). From general properties of Galois correspondences, it follows that a class of representations $\mathcal{X}$ is closed (i.e. $\mathcal{X} = \mathcal{X}^{\alpha\beta}$) if and only if $\mathcal{X} = U^\beta$ for some $U \subseteq KF$. Similarly, a subset U of KF is closed (i.e. $U = U^{\beta\alpha}$) if and only if $U = \mathcal{X}^\alpha$ for some class of representations $\mathcal{X}$. Furthermore, if we restrict the maps α and β to the systems of closed elements, they will be one-to-one and mutually inverse.

We now prove the second assertion of the theorem. By definition, a class of representations $\mathcal{X}$ is a variety if and only if $\mathcal{X} = U^\beta$ for some $U \subseteq KF$. Hence varieties and closed classes are just the same.

Let U be a closed subset in KF. Then $U = \mathcal{X}^\alpha$ for some class $\mathcal{X}$, that is, U is the set of all identities of $\mathcal{X}$. It is easy to see that such a set must be a fully invariant ideal of KF. Conversely, suppose I is a fully invariant ideal of KF and prove that it is closed, that is $I^{\beta\alpha} = I$. Consider first the regular representation Reg $F = (KF, F)$ of F and its factor-representation $\phi = (KF/I, F)$. We will show that the set of all identities of the latter coincides with I.

Let $u(x_1, \ldots, x_n) \in I$. Since I is fully invariant, for any $f_1, \ldots, f_n \in F$ we have $u(f_1, \ldots, f_n) \in I$, and so $u(f_1, \ldots, f_n)$ annihilates the module KF/I. Hence $u(x_1, \ldots, x_n)$ is an identity of the representation ϕ. On the other hand, let $u(x_1, \ldots, x_n)$ be an identity of $\phi = (KF/I, F)$. Take in KF/I the element $1 + I$; since

$$0 = (1 + I) \circ u(x_1, \ldots, x_n) = u(x_1, \ldots, x_n) + I,$$

we see that $u(x_1, \ldots, x_n)$ must belong to I.

In particular, we have $\phi \in I^\beta$. Let now $u \in I^{\beta\alpha}$. This means that u is an identity of the class I^β, so it is satisfied in ϕ. By the above, $u \in I$. Thus $I = I^{\beta\alpha}$. $\square$

For any variety $\mathcal{X}$ of group representations, the ideal $\mathcal{X}^\alpha$ of its identities is denoted by Id $\mathcal{X}$. By Theorem 0.2.2, the map $\mathcal{X} \mapsto \text{Id}\,\mathcal{X}$ is a bijection between the varieties of group representations over K and the fully invariant ideals of KF. The set of varieties of group representations over K is denoted by $\mathbb{M}(K)$. In view of the preceding remark, the behavior of this set is controlled by the free group algebra KF.

Examples. 1. The class $\mathcal{S}$ of all trivial representations (recall that a representation $\rho = (V, G)$ is called trivial if each $g \in G$ acts identically on V) is a variety, for it can be determined by a single identity

$$y \circ (x - 1) \equiv 0.$$

It is easy to see that the ideal Id $\mathcal{S}$ of identities of $\mathcal{S}$ is precisely the augmentation ideal Δ of KF.

2. A representation $\rho = (V, G)$ is called *stable of class n*, or simply *n-stable* (this terminology goes back to Kaloujnine [38]) if there is a series of G-modules

$$0 = A_0 \subseteq A_1 \subseteq \ldots \subseteq A_n = V$$

such that G acts trivially on every factor A_{i+1}/A_i.[1] A typical example of an n-stable representation is the representation $\mathrm{ut}_n(K) = (K^n, \mathrm{UT}_n(K))$ where K^n is the free K-module of rank n and $\mathrm{UT}_n(K)$ is the unitriangular matrix group of degree n over K acting on K^n in the natural way. The class S^n of all n-stable representations is a variety because it is determined by the identity

$$y \circ (x_1 - 1)(x_2 - 1)\ldots(x_n - 1) \equiv 0 .$$

A straightforward verification shows that $\mathrm{Id}(S^n) = \Delta^n$.

3. A representation is said to be n-unipotent if it satisfies the identity

$$y \circ (x - 1)^n \equiv 0.$$

The variety of all n-unipotent representations is denoted by $\mathcal{U}_n$. Evidently $S^n \subseteq \mathcal{U}_n$. On the other hand, suppose that K is a field, then a classical theorem of Kolchin [42] states that if a *finite-dimensional* representation $\rho = (V, G)$ is unipotent, then it is stable (there is no need to speak here about the class of stability and the class of unipotency because they both can be chosen to coincide with $\dim V$). In other words, in the finite-dimensional case stable representations and unipotent representations are exactly the same.

The Kolchin Theorem led naturally to the problem of whether every (not necessarily finite-dimensional) unipotent representation over a field is stable, i.e. whether $\mathcal{U}_n \subseteq S^N$ for some $N = N(n)$. If the ground field K is of prime characteristic, a negative answer can be obtained immediately. But for $\mathrm{char}\, K = 0$ the problem has remained unsolved for about 35 years. We will return to this question in § 1.6.

4. For any variety of groups Θ denote by $\omega\Theta$ the class of all representations $\rho = (V, G)$ such that $G/\mathrm{Ker}\,\rho \in \Theta$. This class is a variety because if Θ is determined by a set of group identities $\{f_i\}$, then $\omega\Theta$ is determined by the set $\{f_i - 1\}$. Note that the map $\Theta \mapsto \omega\Theta$ is injective, for if $G \in \Theta_1 \setminus \Theta_2$, then it is clear that $\mathrm{Reg}\, G = (KG, G) \in \omega\Theta_1 \setminus \omega\Theta_2$. Thus there exists a

---

[1] The word "stable" is overused in today's mathematics and, probably, is not optimal here. But this term is already quite common in the field, so we decided not to change it.

natural embedding of the set of varieties of groups into the set $\mathbb{M}(K)$. The varieties of group representations of the form $\omega\Theta$ will be sometimes called the *varieties of group type*.

It is not hard to identify the ideal $\mathrm{Id}(\omega\Theta)$ of identities of $\omega\Theta$. It is generated, as a right ideal, by all $f - 1$, where f belongs to the Θ-verbal subgroup $\Theta(F)$ of F. Equivalently, $\mathrm{Id}(\omega\Theta)$ is the kernel of the natural epimorphism $KF \to K[F/\Theta(F)]$.

5. This example complements the preceding one and demonstrates that there is an essential difference between identities of abstract groups and identities of their representations. Take the special linear group $\mathrm{SL}_2(K)$ over a field K of characteristic zero. Since $\mathrm{SL}_2(K)$ contains free nonabelian subgroups, it has no nontrivial group identities. Therefore from the standpoint of group identities the classical group $\mathrm{SL}_2(K)$ is not an "interesting" object. Consider now another classical object: the canonical two-dimensional representation $\mathrm{sl}_2(K) = (K^2, \mathrm{SL}_2(K))$. This representation has many interesting identities, for example, an elementary verification shows that

$$(x_1 + x_1^{-1})x_2 - x_2(x_1 + x_1^{-1}) \tag{1}$$

is an identity of $\mathrm{sl}_2(K)$.[2]

A similar phenomenon is valid for other classical matrix groups over an infinite field: as abstract groups, they usually have no nontrivial identities, while their canonical representations certainly do (for instance, by the Amitsur–Levitzki Theorem, every n-dimensional representation satisfies the so-called standard polynomial identity of degree $2n$).

6. Evidently the class $\mathcal{O}$ of all representations over K and the class $\mathcal{E}$ of all zero representations are varieties; they are called *trivial* varieties. The corresponding fully invariant ideals in KF are $\{0\}$ and KF respectively.

0.2.3. Proposition. *If K is a field, then every proper (i.e. $\neq KF$) fully invariant ideal of KF is contained in the augmentation ideal Δ.*

---

[2] Moreover, it was proved in [51] that every identity of $\mathrm{sl}_2(K)$ is a consequence of (1).

Proof. Let I be such an ideal and $u(x_1, \ldots, x_n) = \sum \lambda_i f_i(x_1, \ldots, x_n)$ an element from I. Since I is completely invariant,

$$u(1, \ldots, 1) = \sum \lambda_i f_i(1, \ldots, 1) = \sum \lambda_i \in I.$$

If $\sum \lambda_i \neq 0$, then $1 \in I$, that is $I = KF$, which is impossible. Hence $\sum \lambda_i = 0$, and so $I \subseteq \Delta$. $\square$

Equivalently: *Every nonzero (i.e. $\neq \mathcal{E}$) variety of group representations over a field contains S.*

Let $\mathcal{X}$ be a variety and $\rho = (V, G)$ an arbitrary representation. It is easy to see that if A_i $(i \in I)$ are G-submodules of V such that the corresponding factor-representations $(V/A_i, G)$ belong to $\mathcal{X}$, then $(V/ \cap A_i, G)$ belongs to $\mathcal{X}$ as well. Therefore there exists the smallest G-submodule A of V such that $(V/A, G) \in \mathcal{X}$. This submodule is called the $\mathcal{X}$-*verbal* of ρ and is denoted by $\mathcal{X}^*(\rho) = \mathcal{X}^*(V, G)$.

0.2.4. Lemma. *Let $\mathcal{X}$ be a variety. Then the following assertions are valid:*

(i) *If $\rho = \prod \rho_i$, then $\mathcal{X}^*(\rho) = \prod \mathcal{X}^*(\rho_i)$.*

(ii) *If $\rho \subseteq \sigma$, then $\mathcal{X}^*(\rho) \subseteq \mathcal{X}^*(\sigma)$.*

(iii) *If $\mu : \rho \to \sigma$ is a homomorphism, then $\mathcal{X}^*(\rho^\mu) = (\mathcal{X}^*(\rho))^\mu$.*

Proof is routine. For example, let us prove the last assertion. Recall that two representations are said to be equivalent (see Section 0.1) if their faithful images are isomorphic. It is evident that every representation has the same identities as its faithful image, and therefore if a representation ϕ belongs to a variety $\mathcal{X}$, then all representations equivalent to ϕ belong to $\mathcal{X}$ as well.

Now let $\rho = (V, G)$, $\sigma = (W, H)$, and let $\mu : \rho \to \sigma$ be an epimorphism. Denote $\mathcal{X}^*(\rho) = A$ and $\mathcal{X}^*(\sigma) = B$. We will show that $A^\mu = B$. Note first that the epimorphism μ induces an epimorphism of factor-representations

$$(V/A, G) \to (W/A^\mu, H).$$

Since $(V/A, G) \in \mathcal{X}$ and since every variety is closed under taking epimorphic images, we have $(W/A^\mu, H) \in \mathcal{X}$, whence $A^\mu \supseteq B$. On the other hand, the representations

$$(V/B^{\mu^{-1}}, G) \quad \text{and} \quad (W/B, H)$$

are certainly equivalent, and by the previous remark $(V/B^{\mu^{-1}}, G) \in \mathcal{X}$. Hence $A \subseteq B^{\mu^{-1}}$ and $A^\mu \subseteq B$. $\square$

FREE REPRESENTATIONS. Given two non-empty sets X and Y, take the free group F_X on X and the free KF_X-module $\Phi_Y = \bigoplus_{y \in Y} yKF_X$, having the set Y as a basis. The corresponding representation $\phi = (\Phi_Y, F_X)$ is called the *(absolutely) free representation on a pair of sets* $\{Y, X\}$. This terminology is justified by the following obvious observation: if $\rho = (V, G)$ is any representation, then an arbitrary pair of maps $X \to G$ and $Y \to V$ can be uniquely extended to a homomorphism from ϕ to ρ. Thus ϕ is a free object in the category REP-K. Note, in particular, that the regular representation Reg $F_X = (KF_X, F_X)$ is free. It is called the *cyclic* free representation on the set X.

Now let $\mathcal{X}$ be a variety and let $E_Y = \Phi_Y/\mathcal{X}^*(\phi)$. There naturally arises a representation (E_Y, F_X) called the *free representation of the variety* $\mathcal{X}$ on $\{Y, X\}$. This terminology is also natural because (E_Y, F_X) is a free object of the variety $\mathcal{X}$ regarded as a category.

The *rank* of the representation (E_Y, F_X) is the pair of cardinalities $|Y|$ and $|X|$. If $|Y| = 1$, then (E_Y, F_X) is called the *free cyclic representation of rank* $|X|$ *in* $\mathcal{X}$. The most important is the free cyclic representation of *countable* rank in $\mathcal{X}$ which is denoted by Fr $\mathcal{X}$. According to the above, it is constructed as follows. Take the free group of countable rank F and its regular representation Reg $F = (KF, F)$, then Fr $\mathcal{X}$ is just the factor-representation

$$(KF/\mathcal{X}^*(\text{Reg } F), F).$$

Similarly, the free cyclic representation of a finite rank n in $\mathcal{X}$ is isomorphic to

$$(KF_n/\mathcal{X}^*(\text{Reg } F_n), F_n),$$

F_n being the free group of rank n. This representation is denoted by $\text{Fr}_n\mathcal{X}$.

0.2.5. Proposition. $\mathcal{X}^*(\operatorname{Reg} F) = \operatorname{Id} \mathcal{X}$ *for every variety* $\mathcal{X}$. *Therefore*

$$\operatorname{Fr} \mathcal{X} = (KF/\operatorname{Id} \mathcal{X}, F).$$

P r o o f. Denote $\mathcal{X}^*(\operatorname{Reg} F) = A$, $\operatorname{Id} \mathcal{X} = I$. It is evident that $(KF/I, F) \in \mathcal{X}$ whence $A \subseteq I$. Now let $u(x_1, \ldots, x_n) \in I$. Since $(KF/A, F) \in \mathcal{X}$, u is an identity of this representation. Therefore for the element $\bar{1} = 1 + A \in KF/A$ we have

$$0 = \bar{1} \circ u(x_1, \ldots, x_n) = u(x_1, \ldots, x_n) + A,$$

so that $u(x_1, \ldots, x_n) \in A$. $\square$

Example. To illustrate the above notions, we describe the free cyclic representation $\operatorname{Fr}(\omega\Theta)$ of the variety $\omega\Theta$, Θ being any variety of groups. By the preceding statement, $\operatorname{Fr}(\omega\Theta) = (KF/\operatorname{Id}(\omega\Theta), F)$. Denote by H the Θ-verbal subgroup of F, then $F/H = F(\Theta)$ is the free group of countable rank of Θ. It was earlier noted that $\operatorname{Id}(\omega\Theta)$ coincides with the kernel of the canonical epimorphism $KF \to K[F(\Theta)]$, and so

$$\operatorname{Fr}(\omega\Theta) = (K[F(\Theta)], F)$$

where F acts on $K[F(\Theta)]$ by the natural rule

$$\left(\sum \lambda_i(f_i H)\right) \circ f = \sum \lambda_i(f_i f H).$$

Further, the kernel of $\operatorname{Fr}(\omega\Theta)$ coincides with H, whence the faithful image of this representation is precisely the regular representation of $F(\Theta)$:

$$\overline{\operatorname{Fr}}(\omega\Theta) = \operatorname{Reg} F(\Theta) = (K[F(\Theta)], F(\Theta)).$$

VARIETIES AND CLOSURE OPERATIONS. Since the intersection of any set of varieties is a variety, one can speak of the variety generated by an arbitrary class $\mathcal{K}$ of representations. This variety is denoted by $\operatorname{var} \mathcal{K}$. For example, it is clear that every variety $\mathcal{X}$ is generated by its free cyclic representation of countable rank:

$$\mathcal{X} = \operatorname{var} \{\operatorname{Fr} \mathcal{X}\}.$$

0.2.6. Lemma. *If $\mathcal{K}$ is a class of representations, then all free representations of* var $\mathcal{K}$ *belong to the class* VSC$\mathcal{K}$.

Proof. Denote $\mathcal{X} = $ var $\mathcal{K}$ and $I = $ Id $\mathcal{X}$. We will prove, for instance, that Fr $\mathcal{X} = (KF/I, F) \in $ VSC$\mathcal{K}$. Take the absolutely free representation Reg $F = (KF, F)$ and let μ_α, $\alpha \in M$, be all possible homomorphisms from Reg F to $\mathcal{K}$-representations. For each $\alpha \in M$ denote the kernels of μ_α in KF and in F by A_α and G_α respectively. Then the representation $(KF/A_\alpha, F/G_\alpha)$ is isomorphically embedded in some representation from $\mathcal{K}$. Let $A = \cap A_\alpha$ and $G = \cap G_\alpha$. Using the Remak Theorem, we conclude that

$$(KF/A,\ F/G) \in \text{SC}\mathcal{K}$$

whence $(KF/A, F) \in $ VSC$\mathcal{K}$. It remains to show that $A = I$.

Since $(KF/A, F) \in \mathcal{X}$, we have $I \subseteq A$. To prove the reverse inclusion, take $u = u(x_1, \ldots, x_n) \in A$. Since $\mathcal{X} = $ var $\mathcal{K}$, the ideal I consists of all identities of the class $\mathcal{K}$, so that it is enough to show that u is an identity of any representation $\rho = (B, H)$ from $\mathcal{K}$. Assume the contrary:

$$\exists b \in B,\ h_1, \ldots, h_n \in H: \quad b \circ u(h_1, \ldots, h_n) \neq 0.$$

Consider the homomorphism $\mu : (KF, F) \to (B, H)$ defined as follows: $1^\mu = b$ where 1 is the unit of KF; $x_i^\mu = h_i$ $(i = 1, \ldots, n)$ and $x_i^\mu = x_i$ for $i > n$. Since A is the intersection of the "left" kernels of all such homomorphisms and $u \in A$, we have $u^\mu = 0$. But

$$u^\mu = (1 \circ u)^\mu = b \circ u(h_1, \ldots, h_n) \neq 0.$$

This contradiction shows that $u \in I$. $\square$

It is now easy to prove an analogue of the classical Birkhoff Theorem [7] giving a characterization of varieties in terms of closure operations on classes of representations.

0.2.7. Theorem. *A class of group representations is a variety if and only if it is closed with respect to the operations* V, Q, S, C.

Proof. Evidently every variety is closed with respect to these operations. Conversely, suppose $\mathcal{X} = V\mathcal{X} = Q\mathcal{X} = S\mathcal{X} = C\mathcal{X}$. By 0.2.6, all free representations of var $\mathcal{X}$ are contained in VSC$\mathcal{X} = \mathcal{X}$. Since every representation from var $\mathcal{X}$ is an epimorphic image of some free representation of var $\mathcal{X}$, it follows that var $\mathcal{X} = \mathcal{X}$. □

Using this theorem and easily verified relations between closure operations (e.g. QV $\leq$ VQ, SQ $\leq$ QS, etc.), we obtain:

0.2.8. Corollary. var $\mathcal{K} = $ VQSC$\mathcal{K}$ *for any class* $\mathcal{K}$. □

Note. Since group representations are *two-sorted* algebraic structures, one should not be surprised when some statements on varieties of group representations do not repeat literally the well known properties of varieties of "usual" algebras. For example, the statement of Theorem 0.2.7 differs from that of the Birkhoff Theorem (the former contains an additional requirement of V-closedness). However, it is easy to understand that this difference is immaterial. Indeed, if we consider any class of group representations $\mathcal{X}$ and the corresponding class V$\mathcal{X}$, then from the standpoint of identities there is no difference between $\mathcal{X}$ and V$\mathcal{X}$ at all: both classes have the same identities. (In particular, this shows that every variety of group representations simply *must* be V-closed).

0.3. Certain properties of varieties.

THE ALGEBRA OF VARIETIES. For arbitrary varieties $\mathcal{X}$ and $\mathcal{Y}$ their *product* $\mathcal{X}\mathcal{Y}$ is the class of all representations $\rho = (V, G)$ such that V has a G-submodule A with $(A, G) \in \mathcal{X}$ and $(V/A, G) \in \mathcal{Y}$. A straightforward verification shows that $\mathcal{X}\mathcal{Y}$ is also a variety and

$$\text{Id}(\mathcal{X}\mathcal{Y}) = \text{Id}\,\mathcal{Y} \cdot \text{Id}\,\mathcal{X}.$$

Therefore under this multiplication the set $\mathbb{M}(K)$ of varieties of group representations over a given K forms a *semigroup* which is anti-isomorphic to the semigroup of fully invariant ideals of KF. The varieties $\mathcal{O}$ and $\mathcal{E}$ are respectively 0 and 1 of the semigroup $\mathbb{M}(K)$.

For arbitrary varieties $\mathcal{X}$ and $\mathcal{Y}$ their *join* $\mathcal{X} \vee \mathcal{Y}$ and *meet* $\mathcal{X} \wedge \mathcal{Y}$ are defined in the standard way:

$$\mathcal{X} \vee \mathcal{Y} = \mathrm{var}\,(\mathcal{X} \cup \mathcal{Y}), \quad \mathcal{X} \wedge \mathcal{Y} = \mathcal{X} \cap \mathcal{Y} \,.$$

The set of varieties of group representations over a given K is a *complete lattice* under the operations $\vee$ and $\wedge$. Since

$$\mathrm{Id}\,(\mathcal{X} \vee \mathcal{Y}) = \mathrm{Id}\,\mathcal{X} \cap \mathrm{Id}\,\mathcal{Y}, \quad \mathrm{Id}\,(\mathcal{X} \wedge \mathcal{Y}) = \mathrm{Id}\,\mathcal{X} + \mathrm{Id}\,\mathcal{Y},$$

this lattice is dual to the lattice of completely invariant ideals of KF.

Now let $\mathcal{X}$ be a variety of representations, and Θ a variety of groups. Denote by $\mathcal{X} \times \Theta$ the class of all representations (V, G) such that G has a normal subgroup H with $(V, H) \in \mathcal{X}$ and $G/H \in \Theta$. It is easy to show (for instance, using 0.2.7) that $\mathcal{X} \times \Theta$ is a variety of group representations. In particular, the variety $\omega\Theta$ can now be redefined as $\omega\Theta = \mathcal{S} \times \Theta$, where $\mathcal{S}$ is the variety of trivial representations.

Recall that the product $\Theta_1\Theta_2$ of two varieties of groups Θ_1 and Θ_2 is the class of all groups G having a normal subgroup A with $A \in \Theta_1$ and $G/A \in \Theta_2$. It is known that $\Theta_1\Theta_2$ is a variety of groups and that this multiplication is associative, so one can speak of the *semigroup of varieties of groups*. The reader interested in the structure of this semigroup, is referred to Chapter 2 of [68].

A direct verification shows that

$$(\mathcal{X}\mathcal{Y}) \times \Theta = (\mathcal{X} \times \Theta)(\mathcal{Y} \times \Theta), \quad (\mathcal{X} \times \Theta_1) \times \Theta_2 = \mathcal{X} \times (\Theta_1\Theta_2),$$

where $\mathcal{X}$ and $\mathcal{Y}$ are varieties of representations, Θ_1 and Θ_2 are varieties of groups. In other words, $\times$ is an *action* of the semigroup of varieties of groups on the semigroup of varieties of group representations.

FINITELY BASED AND SPECHT VARIETIES.   A variety is said to be *finitely based* if it can be determined by a finite set of identities. A variety $\mathcal{X}$ is said to be *Specht* if all its subvarieties (including $\mathcal{X}$) are finitely based. One can easily show that a variety is Specht if and only if it is finitely based and satisfies the descending chain condition on subvarieties. Indeed, a variety $\mathcal{X}$ is Specht if and only if all verbal overideals of $\operatorname{Id}\mathcal{X}$ in KF (including $\operatorname{Id}\mathcal{X}$) are finitely generated as fully invariant ideals. By a standard Noetherian-type argument, this is equivalent to the ascending chain conditions on these overideals which, in turn, is equivalent to the descending chain condition on subvarieties of $\mathcal{X}$.

0.3.1. Proposition. *If $\mathcal{X}$ and $\mathcal{Y}$ are finitely based, then $\mathcal{X}\mathcal{Y}$ is also finitely based. Therefore finitely based varieties form a subsemigroup in $\mathbb{M}(K)$.*

Proof. Let $\operatorname{Id}\mathcal{X} = I$, $\operatorname{Id}\mathcal{Y} = J$. Since $\mathcal{X}$ and $\mathcal{Y}$ are finitely based, I and J are finitely generated as fully invariant ideals of KF. For any subset M of KF denote by $\operatorname{Id}(M)$ the fully invariant ideal of KF generated by M. Then, without loss of generality, we may assume that $I = \operatorname{Id}(v)$, $J = \operatorname{Id}(u)$ for some $u, v \in KF$ and, moreover, that the set of free generators x_i ocurring in u does not intersect with the analogous set for v. We will prove that $JI = \operatorname{Id}(uv)$.

Let $u = u(x_1, \ldots, x_m)$, $v = v(x_{m+1}, \ldots, x_{m+n})$. Every element from JI is a K-linear combination of elements

$$au(f_1, \ldots, f_m)bv(g_1, \ldots, g_n)c, \quad \text{where } a, b, c, f_i, g_i \in F.$$

Since $u(f_1, \ldots, f_m)b = bu(f_1^b, \ldots, f_m^b)$ where $f_i^b = b^{-1}f_i b$, every element from JI is a K-linear combination of elements of the form

$$au(f_1, \ldots, f_m)v(g_1, \ldots, g_n)c, \quad \text{where } a, c, f_i, g_i \in F.$$

But it is evident that these elements belong to $\operatorname{Id}(uv)$. Therefore $JI = \operatorname{Id}(uv)$ and, since $JI = \operatorname{Id}(\mathcal{X}\mathcal{Y})$, $\mathcal{X}\mathcal{Y}$ is finitely based. $\square$

The following question is still unanswered.

0.3.2. Problem. *Is the product of two Specht varieties of group representations also a Specht variety?*

It is unlikely that the answer here is positive (even over a field), although we do not have any definite arguments.

0.3.3. Proposition. *A variety $\omega\Theta$ is finitely based if and only if Θ is a finitely based variety of groups.*

P r o o f. Let Θ be determined by identities $f_i \equiv 1$, $i \in I$, then $\omega\Theta$ is determined by the identities $f_i - 1$, $i \in I$. Therefore if Θ is finitely based, so is $\omega\Theta$. Conversely, let $\omega\Theta$ be finitely based. This means that $I = \mathrm{Id}(\omega\Theta)$ can be generated, as a verbal ideal, by a finite set of elements $u_1, \ldots, u_m$. On the other hand, I is generated, as a verbal ideal, by the elements $f_i - 1$. Therefore each u_k is a finite sum of elements of the form $a(f_i - 1)^{\varphi}b$, where $a, b \in KF$, $\varphi \in \mathrm{End}\, F$. Hence I is generated, as a verbal ideal, by a *finite* number of the $f_i - 1$, say by $f_1 - 1, \ldots, f_n - 1$. But then it is easy to see that Θ is determined by the identities $f_1 \equiv 1, \ldots, f_n \equiv 1$. Indeed, if they are satisfied in some group G, then $\mathrm{Reg}_K G$ satisfies the identities $f_1 - 1, \ldots, f_n - 1$. Hence $\mathrm{Reg}_K G \in \omega\Theta$ and, since this representation is faithful, $G \in \Theta$. $\square$

In particular, Proposition 0.3.3 guarantees that over an arbitrary K there exist varieties of group representations which are not finitely based (because non-finitely based varieties of groups do exist [70], [1], [88]).

THE AXIOMATIC RANK AND THE BASIC RANK. As usual, F and F_n are free groups of countable rank and of a finite rank n respectively. Consider two absolutely free representations

$$\mathrm{Reg}\, F = (KF, F) \quad \text{and} \quad \mathrm{Reg}\, F_n = (KF_n, F_n),$$

and let us suppose that $\mathrm{Reg}\, F_n$ is naturally embedded in $\mathrm{Reg}\, F$. Take an arbitrary variety $\mathcal{X}$ and let $I = \mathcal{X}^*(\mathrm{Reg}\, F) = \mathrm{Id}\, \mathcal{X}$ be the $\mathcal{X}$-verbal ideal of KF, but $I_n = \mathcal{X}^*(\mathrm{Reg}\, F_n)$ the corresponding ideal of KF_n. The elements of I_n are called the *identities in n variables* of $\mathcal{X}$. Recall (see page 11) that

$$\mathrm{Fr}\, \mathcal{X} = (KF/I, F) \quad \text{and} \quad \mathrm{Fr}_n \mathcal{X} = (KF_n/I_n, F_n)$$

are the free cyclic representations in $\mathcal{X}$ of countable rank and of rank n respectively.

The next three lemmas are straightforward. We will prove only the first of them, leaving the remaining ones to the reader.

0.3.4. Lemma. $I_n = I \cap K F_n$.

P r o o f. By 0.2.4 (ii), $\mathcal{X}^*(\operatorname{Reg} F_n) \subseteq \mathcal{X}^*(\operatorname{Reg} F)$, whence $I_n \subseteq I \cap K F_n$. On the other hand, consider the epimorphism $\mu : \operatorname{Reg} F \to \operatorname{Reg} F_n$ defined by the rule

$$x_i^\mu = x_i \qquad \text{if} \qquad i = 1, \ldots, n;$$
$$x_i^\mu = 1 \qquad \text{if} \qquad i \geq n.$$

Then $I \cap K F_n = (I \cap K F_n)^\mu \subseteq I^\mu$ and, by 0.2.4 (iii),

$$I^\mu = (\mathcal{X}^*(\operatorname{Reg} F_n))^\mu = \mathcal{X}^*(\operatorname{Reg} F_n) = I_n.$$

Therefore $I \cap K F_n \subseteq I_n$. $\square$

0.3.5. Lemma. *A representation of an n-generated group belongs to* $\mathcal{X}$ *if and only if this representation satisfies the identities in n variables of* $\mathcal{X}$. $\square$

For any variety $\mathcal{X}$ and any positive integer n, the variety determined by the identities in n variables of $\mathcal{X}$ is denoted by $\mathcal{X}^{(n)}$.

0.3.6. Lemma. $\mathcal{X}^{(n)}$ *consists of all representations* (V, G) *such that if* H *is an arbitrary n-generated subgroup of* G, *then* $(V, H) \in \mathcal{X}$. $\square$

Thus, for every variety $\mathcal{X}$ we have a descending chain of overvarieties

$$\mathcal{X}^{(1)} \supseteq \mathcal{X}^{(2)} \supseteq \cdots \supseteq \mathcal{X}^{(n)} \supseteq \cdots \supseteq \mathcal{X} = \bigcap_{n=1}^{\infty} \mathcal{X}^{(n)}. \qquad (1)$$

If there exists n such that $\mathcal{X} = \mathcal{X}^{(n)}$, then the first n with this property is called the *axiomatic rank* of $\mathcal{X}$ and is denoted by $r_a(\mathcal{X})$. Otherwise $\mathcal{X}$ is called a variety of infinite axiomatic rank.

On the other hand, if we denote $\mathcal{X}_{(n)} = \mathrm{var}\,(\mathrm{Fr}_n \mathcal{X})$, then there arises an ascending chain of varieties

$$\mathcal{X}_{(1)} \subseteq \mathcal{X}_{(2)} \subseteq \cdots \subseteq \mathcal{X}_{(n)} \subseteq \cdots \subseteq \mathcal{X} = \bigvee_{n=1}^{\infty} \mathcal{X}_{(n)}. \tag{2}$$

If $\mathcal{X} = \mathcal{X}_{(n)}$ for some n, then the first such n is called the *basis rank* of $\mathcal{X}$ and is denoted by $r_b(\mathcal{X})$. If $\mathcal{X}_{(n)} < \mathcal{X}$ for every n, we say that $r_b(\mathcal{X}) = \infty$.

Examples. 1. Let Θ be a variety of groups which cannot be determined by identities in a finite number of variables (such varieties were first constructed by Vaughan-Lee [88]). Then $r_a(\omega\Theta) = \infty$.

2. Let K be a field of characteristic p, and let $\mathcal{A}_p$ be the variety of abelian groups of exponent p. It is easy to show that the variety $\omega\mathcal{A}_p$ is not stable, while every finitely generated representation from $\omega\mathcal{A}_p$ is stable. Hence $\omega\mathcal{A}_p$ cannot be generated by a finitely generated representation, and so $r_b(\omega\mathcal{A}_p) = \infty$.

3. Let K be a field of characteristic 0, $\mathcal{A}$ the variety of all abelian groups and $\mathcal{X}$ an arbitrary subvariety of $\omega\mathcal{A}$. Then $r_a(\mathcal{X}) = 1$ [49].

4. Evidently $r_a(\mathcal{S}) = r_b(\mathcal{S}) = 1$. More generally, it will be proved in §1.1 that if a variety $\mathcal{X}$ is n-stable, then $r_a(\mathcal{X}) \leq n$ and $r_b(\mathcal{X}) \leq n$.

0.4. Changing the ground ring

Recall that $\mathbb{M}(K)$ denotes the set of varieties of group representations over K. According to the previous section, $\mathbb{M}(K)$ is a semigroup under the multiplication of varieties.

Let K be a commutative ring with 1 and let R be its subring (with 1). In the present section, our objective is to describe certain connections, discovered by Plotkin [78], between varieties of group representations over K and over R. Consider two maps

$$\nu : \ \mathbb{M}(R) \to \mathbb{M}(K) \qquad \text{and} \qquad \nu' : \ \mathbb{M}(K) \to \mathbb{M}(R)$$

defined as follows. If $\mathcal{X}$ is a variety of representations over R, then $\mathcal{X}^\nu$ is the class of all representations over K which, regarded as representations over R, belong to $\mathcal{X}$. Clearly $\mathcal{X}^\nu$ is a variety of representations over K (for example, by Theorem 0.2.7). Conversely, let $\mathcal{Y}$ be a variety of representations over K. Consider $\mathcal{Y}$ as a class of representations over R and set $\mathcal{Y}^{\nu'} = \mathrm{var}_R \mathcal{Y}$.

It follows from the definitions that $\mathcal{X}^{\nu\nu'} \subseteq \mathcal{X}$ and $\mathcal{Y}^{\nu'\nu} \supseteq \mathcal{Y}$ for each $\mathcal{X} \in \mathcal{M}(R)$ and $\mathcal{Y} \in \mathcal{M}(K)$. In general, these inclusions are strict.

Examples. 1. Let $R = \mathbb{Z}$ and $K = \mathbb{Q}$, and let $\mathcal{X}$ be the variety over $\mathbb{Z}$ defined by the identity $n(x - 1)$ with $n > 1$. Then the inclusion $\mathcal{X}^{\nu\nu'} \subseteq \mathcal{X}$ is strict. Indeed, $\mathcal{X}$ consists of all representations $\rho = (V, G)$ over $\mathbb{Z}$ where V has a G-submodule A such that A is an abelian group of exponent n and G acts trivially on V/A. From the definition of ν, it follows that $\mathcal{X}^\nu$ is the variety of trivial representations over $\mathbb{Q}$, whence $\mathcal{X}^{\nu\nu'} = S$. But it is clear that $\mathcal{X} \neq S$.

2. Examples showing that the inclusion $\mathcal{Y}^{\nu'\nu} \supseteq \mathcal{Y}$ is, in general, strict are less elementary and will be given later. However, this fact can be immediately deduced from general considerations. Namely, let K be a field whose cardinality is greater than the continuum $\mathfrak{c}$, and let R be a countable subfield. It is known that $|\mathrm{M}(R)| = \mathfrak{c}$, while $|\mathrm{M}(K)| = |K| > \mathfrak{c}$ (see [79] or [80, Theorem 2.5.1]). Therefore the map ν' is not injective and, *a fortiori*, the equality $\mathcal{Y}^{\nu'\nu} = \mathcal{Y}$ is false for some $\mathcal{Y} \in \mathrm{M}(K)$.

It is not hard to understand how the ideals of identities of $\mathcal{X}^\nu$ and $\mathcal{Y}^{\nu'}$ depend on those of $\mathcal{X}$ and $\mathcal{Y}$. Indeed, suppose that RF is naturally contained in KF and let $I = \mathrm{Id}\,\mathcal{X}$ and $J = \mathrm{Id}\,\mathcal{Y}$. Denote by KI the ideal of KF generated by I. Then a direct verification shows that

$$\mathrm{Id}\,(\mathcal{X}^\nu) = KI \qquad and \qquad \mathrm{Id}\,(\mathcal{Y}^{\nu'}) = J \cap RF.$$

In particular, since $K(I_1 I_2) = (KI_1)(KI_2)$, we obtain:

0.4.1. Proposition. *The map $\nu : \mathrm{M}(R) \to \mathrm{M}(K)$ is a homomorphism of semigroups.* $\square$

0.4.2. Proposition. *If R and K are fields, then $\mathcal{X}^{\nu\nu'} = \mathcal{X}$ for every $\mathcal{X} \in \mathrm{M}(R)$. In particular, ν is a monomorphism.*

P r o o f. To establish the nontrivial part, take an arbitrary representation $\rho = (V, G)$ from $\mathcal{X}$. Consider the vector K-space $V_K = K \otimes_R V$. The group G acts on V_K by the rule $(\lambda \otimes v) \circ g = \lambda(v \circ g)$, so that there arises a representation $\rho_K = (V_K, G)$. It is clear that ρ_K, as a representation over R, belongs to $\mathcal{X}$. Therefore as a K-representation it belongs to $\mathcal{X}^\nu$. But ρ is a R-subrepresentation of ρ_K, and so $\rho \in \mathcal{X}^{\nu\nu'}$. Hence $\mathcal{X} \subseteq \mathcal{X}^{\nu\nu'}$. $\square$

The remaining results of this section show that in certain cases the investigation of varieties over a given field K can be "localized". More exactly, one can reduce it to studying varieties over finitely generated subrings of K, and thereafter to studying varieties over certain finite fields.

Let K be a commutative ring with identity and let $\{R_i | i \in I\}$ be the system of all finitely generated subrings of K. For every $i \in I$ the embedding $R_i \subseteq K$ determines the maps

$$\nu : \ \mathbb{M}(R_i) \to \mathbb{M}(K) \qquad \text{and} \qquad \nu' : \ \mathbb{M}(K) \to \mathbb{M}(R_i)$$

which will be denoted below by ν_i and ν_i' respectively. If now $\mathcal{Y}$ is a variety of group representations over K, then we set $\mathcal{Y}^{\nu_i'} = \mathcal{X}_i$. In other words, $\mathcal{X}_i$ is a variety of group representations over R_i which is generated by the class $\mathcal{Y}$ regarded as a class of R_i-representations. Furthermore, let us agree that if ρ is a representation over K, then the symbol ρ_{R_i} means that ρ is regarded as a representation over R_i.

0.4.3. Lemma. *A representation ρ over K belongs to $\mathcal{Y}$ if and only if $\rho_{R_i} \in \mathcal{X}_i$ for every $i \in I$. This representation generates $\mathcal{Y}$ if and only if ρ_{R_i} generates $\mathcal{X}_i$ for every $i \in I$.*

P r o o f. If $\rho \in \mathcal{Y}$, then $\rho_{R_i} \in \mathcal{X}_i$. Conversely, suppose that $\rho_{R_i} \in \mathcal{X}_i$ for every $i \in I$, and prove that $\rho \in \mathcal{Y}$. Let $u \in KF$ be an arbitrary identity of $\mathcal{Y}$. Then $u \in R_i F$ for some R_i, so that u is an identity of the corresponding $\mathcal{X}_i$. Since $\rho_{R_i} \in \mathcal{X}_i$, it follows that u is an identity of ρ_{R_i}, as required.

We prove now the second assertion of the lemma. Let ρ generate $\mathcal{Y}$ over K: $\text{var}_K \rho = \mathcal{Y}$. Then

$$\mathcal{X}_i = \text{var}_{R_i} \mathcal{Y} = \text{var}_{R_i}(\text{var}_K \rho) = \text{var}_{R_i}(\rho_{R_i}).$$

Conversely, suppose that $\text{var}_{R_i}(\rho_{R_i}) = \mathcal{X}_i$ for every $i \in I$. Then, in view of the above, $\rho \in \mathcal{Y}$, and it remains to show that every identity v of ρ is an identity of the variety $\mathcal{Y}$. Choose a finitely generated subring R_i of K such that $v \in R_i F$, then v is an identity of ρ_{R_i} and so it is an identity of $\mathcal{X}_i$. Since $\mathcal{Y} \subseteq \mathcal{X}_i$ (over $R_i!$), v is an identity of $\mathcal{Y}$ as well. $\square$

0.4.4. Lemma. $r_b(\mathcal{Y}) \leq n \iff \forall i \in I : r_b(\mathcal{X}_i) \leq n$.

Proof. Let $J = \text{Id}_n \mathcal{Y}$ be the ideal of identities of $\mathcal{Y}$ in KF_n. If $r_b(\mathcal{Y}) \leq n$, then $\mathcal{Y}$ is generated by the representation $\text{Fr}_n \mathcal{Y} = (KF_n/J, F_n)$. By the previous lemma, $\mathcal{X}_i$ is generated by $(\text{Fr}_n \mathcal{Y})_{R_i}$, whence $r_b(\mathcal{X}_i) \leq n$.

Conversely, let $r_b(\mathcal{X}_i) \leq n$ for every i. Considering $R_i F$ and J as R_i-submodules of KF_n, we have an isomorphism of R_i-modules

$$(R_i F_n + J)/J \simeq R_i F_n/(J \cap R_i F_n). \tag{1}$$

Since $J \cap R_i F_n$ is the ideal of identities of $\mathcal{X}_i$ in $R_i F_n$, we have

$$\text{Fr}_n \mathcal{X}_i = (R_i F_n/(J \cap R_i F_n), \ F_n). \tag{2}$$

By assumption, $\mathcal{X}_i$ is generated by $\text{Fr}_n \mathcal{X}_i$. It follows from (1) and (2) that $\text{Fr}_n \mathcal{X}_i$ is a R_i-subrepresentation of $\text{Fr}_n \mathcal{Y}$. Therefore $\mathcal{X}_i$ is generated by $(\text{Fr}_n \mathcal{Y})_{R_i}$ as well. Since this is valid for arbitrary $i \in I$, it follows from Lemma 0.4.3 that $\text{Fr}_n \mathcal{Y}$ generates $\mathcal{Y}$. $\square$

Let A be a module over a commutative ring R with 1. If $k = R/\text{Ann}_R A$ is a field, A can also be regarded as a vector space over k. In this case we will say that A is an *R-module over the field k*. In a similar sense we speak of an *R-representation over the field k*. Using this terminology and notation, we can now prove the following statement.

0.4.5. Proposition. *Let K be a field of characteristic zero, R a finitely generated subring of K, and $\nu' : \mathbb{M}(K) \to \mathbb{M}(R)$ the corresponding map on varieties. If a variety $\mathcal{Y}$ over K is generated by finite-dimensional representations, then for any infinite set π of primes the variety $\mathcal{Y}^{\nu'}$ over R can be generated by finite-dimensional R-representations over finite fields whose characteristics belong to π.*

Proof. The variety $\mathcal{Y}$ is generated by finite-dimensional representations of finitely generated groups. Let $\rho = (V, G)$ be such a representation. Choose a basis $e_1, \ldots, e_n$ of V and let $g_1, \ldots, g_m$ be a finite generating set of G closed under taking inverse elements. If

$$e_i \circ g_j = \sum \lambda_{ijk} e_k$$

then denote by R_1 the subring of K generated by R and all the λ_{ijk}. Let A be the R_1-submodule of V spanned by $e_1, \ldots, e_n$. Clearly A is a free R_1-module of rank n, invariant under G.

Let π be an infinite set of primes. Being a finitely generated subring of a field of characteristic zero, R_1 can be approximated by finite fields whose characteristics belong to π (see for example [103, Lemma 10.2]). In other words, there is a set of ideals $\{\mathfrak{a}_i\}$ of R_1 such that $\bigcap \mathfrak{a}_i = 0$ and each $R_1/\mathfrak{a}_i$ is a finite field whose characteristic belongs to π. For each i the R_1-submodule $\mathfrak{a}_i A$ of A is invariant under G, the factor $A/\mathfrak{a}_i A$ is a finite-dimensional vector space over $R_1/\mathfrak{a}_i$, and $\bigcap \mathfrak{a}_i A = 0$.

Now let us consider A as an R-module. Since $R_1/\mathfrak{a}_i$ is a finite field, it follows that $R/\mathrm{Ann}_R(A/\mathfrak{a}_i A)$ is also a finite field of the same characteristic. Thus, the R-representation $\sigma = (A, G)$ is approximated by finite-dimensional representations over finite fields whose characteristics belong to π. Therefore σ is contained in the variety generated by all these finite-dimensional representations. Since $\rho = (V, G)$ is the "K-envelope" of $\sigma = (A, G)$, it is easy to see that ρ_R and σ generate the same variety over R. It remains to note that $\mathcal{Y}^{\nu'}$ is generated by all possible ρ_R, ρ being as above. $\square$

0.5. The free group algebra

In the present section we describe several basic properties of the free group algebra KF which will be necessary in what follows. Our presentation is based on the embedding of KF into the algebra of formal power series, going back to Magnus [57], and on the free differential calculus of Fox [17]. All the material of this section is classical and can be found in many sources;

apart from the original papers [57] and [17], one can mention the books [26, 60, 80, 27].

THE FOX DERIVATIONS. As usual, let F be the absolutely free group of countable rank on the alphabet $X = \{x_1, x_2, \dots\}$ and let K be an arbitrary commutative ring with 1. Take the group algebra KF and let $\epsilon : KF \to K$ be the augmentation map. For any $u = u(x_1, \dots, x_n) \in KF$, the scalar $u^\epsilon = u(1, \dots, 1)$ will be briefly denoted by $u(1)$. A map $D : KF \to KF$ is called a (left) *derivation* if it is K-linear and

$$D(uv) = Du \cdot v(1) + u\, Dv \qquad (1)$$

far all $u, v \in KF$. It is easy to see that every derivation D has the following properties:

$$D\lambda = 0 \qquad (\lambda \in K), \qquad (2)$$

$$Df^{-1} = -f^{-1}Df \qquad (f \in F), \qquad (3)$$

$$D(u_1 u_2 \dots u_k) = \sum_{i=1}^{k} u_1 \dots u_{i-1} \cdot Du_i \cdot u_{i+1}(1) \dots u_k(1) \qquad (u_i \in KF). \quad (4)$$

For each positive integer i consider the map $\partial_i : X \to KF$ defined by the rule $\partial_i x_j = \delta_{ij}$, where δ_{ij} is the Kronecker symbol. This map can be extended to F by (2)–(4) and then to KF by linearity. A straightforward verification shows that the resulting map $\partial_i : KF \to KF$ is uniquely determined and is a derivation. It is called the *i-th Fox derivation* and is also denoted by D_i or $\frac{\partial}{\partial x_i}$. Evidently, if x_i is not involved in u, then $\partial_i u = 0$.

The set Der KF of all derivations on KF can be regarded as a right KF-module, if one defines the KF-linear operations on Der KF by

$$(D_1 + D_2)u = D_1 u + D_2 u \qquad \text{and} \qquad (D \cdot v)u = (Du)v.$$

The structure of this module is rather transparent and depends heavily on the Fox derivations ∂_i.

0.5.1. Proposition. *For every sequence $w_1, w_2, \dots, w_n, \dots$ of elements from KF there exists one and only one derivation D on KF such that $Dx_i = w_i$; it is defined by*

$$Du = \sum_i \partial_i u \cdot w_i. \qquad (5)$$

P r o o f . Since $\partial_i u = 0$ for all but a finite number of indices i, this sum is finite. Further, since $\operatorname{Der} KF$ is a KF-module, the map D defined by (5) is a derivation taking x_i to w_i $(i = 1, 2, \ldots)$. Suppose D' is another derivation such that $D' x_i = w_i$, then $D - D'$ is a derivation taking x_i to 0. By (2)–(4), this implies that $(D - D')u = 0$ for every $u \in KF$, whence $D = D'$. $\square$

The following simple but significant fact is commonly known as the *Fox fundamental formula*.

0.5.2. Proposition. *For every $u \in KF$*

$$u = u(1) + \sum_i \partial_i u \cdot (x_i - 1). \qquad (6)$$

P r o o f . It is easy to verify that the map $u \mapsto u - u(1)$ is a derivation of KF taking x_i to $x_i - 1$. By Proposition 0.5.1, it follows that

$$u - u(1) = \sum_i \partial_i u \cdot (x_i - 1). \qquad \square$$

THE FOX DERIVATIONS OF HIGHER DEGREES. For arbitrary positive integers $i_1, i_2, \ldots, i_n$ define the map $\partial_{i_n \ldots i_1} : KF \to KF$ inductively:

$$\partial_{i_n \ldots i_1} u = \partial_{i_n}(\partial_{i_{n-1} \ldots i_1} u).$$

Such maps are called the *Fox derivations of higher degrees*; clearly they all are K-linear.

0.5.3. Proposition. *For every $u \in KF$ and every $n > 0$ the following "Taylor formula with residue" is valid:*

$$u = u(1) + \sum_{i_1}(\partial_{i_1} u(1))(x_{i_1} - 1) + \sum_{i_2, i_1}(\partial_{i_2 i_1} u(1))(x_{i_2} - 1)(x_{i_1} - 1) + \ldots$$

$$+ \sum_{i_{n-1}, \ldots, i_1} (\partial_{i_{n-1} \ldots i_1} u(1))(x_{i_{n-1}} - 1) \ldots (x_{i_1} - 1) + \qquad (7)$$

$$+ \sum_{i_n, \ldots, i_1} (\partial_{i_n \ldots i_1} u)(x_{i_n} - 1) \ldots (x_{i_1} - 1).$$

Proof. Applying (6) repeatedly to $\partial_{i_1} u$, $\partial_{i_2 i_1} u, \ldots$, we obtain

$$\partial_{i_1} u = \partial_{i_1} u(1) + \sum_i (\partial_{ii_1} u)(x_i - 1),$$

$$\partial_{i_2 i_1} u = \partial_{i_2 i_1} u(1) + \sum_i (\partial_{ii_2 i_1} u)(x_i - 1),$$

and so on. Therefore

$$u = u(1) + \sum_{i_1} (\partial_{i_1} u)(x_{i_1} - 1)$$

$$= u(1) + \sum_{i_1} [\partial_{i_1} u(1) + \sum_{i_2} (\partial_{i_2 i_1} u)(x_{i_2} - 1)] (x_{i_1} - 1)$$

$$= u(1) + \sum_{i_1} (\partial_{i_1} u(1))(x_{i_1} - 1) + \sum_{i_2, i_1} (\partial_{i_2 i_1} u)(x_{i_2} - 1)(x_{i_1} - 1)$$

$$= \ldots \quad . \quad \square$$

Thus, the basic formulas of the Fox differential calculus have been obtained. Using these formulas, let us prove now several statements on the augmentation ideal Δ and its powers.

0.5.4. Proposition. *An element $u \in KF$ belongs to Δ^n if and only if*

$$u(1) = \partial_{i_1} u(1) = \partial_{i_2 i_1} u(1) = \cdots = \partial_{i_{n-1} \ldots i_1} u(1) = 0$$

for arbitrary indices $i_1, i_2, \ldots$.

Proof. If the condition is satisfied, then (7) implies

$$u = \sum (\partial_{i_n \ldots i_1} u)(x_{i_n} - 1) \ldots (x_{i_1} - 1)$$

whence $u \in \Delta^n$. Conversely, if $u \in \Delta^n$, then u is a sum of elements of the form $v_1 v_2 \ldots v_n$ where $v_i \in \Delta$. Hence $u(1) = 0$ and, applying induction on m and (4), one can directly calculate that

$$\partial_{i_m \ldots i_1} u(1) = 0 \quad \text{for} \quad m \le n - 1,$$

as required. $\square$

Let $f = x_{i_1}^{\varepsilon_1} x_{i_2}^{\varepsilon_2} \ldots x_{i_k}^{\varepsilon_k}$ ($\varepsilon_i = \pm 1$) be an element from F, written in the irreducible form. The *length* k of f is denoted by $l(f)$. If $u = \sum \lambda_i f_i \in KF$, where $\lambda_i \in K$ and $f_i \in F$, then the length $l(u)$ is defined as

$$l(u) = \max_i l(f_i).$$

By definition, we let $l(\lambda) = 0$ for $\lambda \in K$.

0.5.5. Lemma. $0 \neq u \in \Delta^n \implies l(u) \geq n/2$.

Proof. Induction on n. For $n = 1, 2$ the assertion is obvious. Let $n \geq 3$ and suppose u is an element of Δ^n such that $l = l(u) < n/2$. Each monomial of u ends with some x_j or x_j^{-1}, therefore

$$u = \sum_j (v^{(j)} x_j + w^{(j)} x_j^{-1})$$

where $l(v^{(j)}) < l$ and $l(w^{(j)}) < l$. Since ∂_j does not enlarge the length of words and $\partial_j x_j^{-1} = -x_j^{-1}$, we have

$$\partial_j u = a^{(j)} - w^{(j)} x_j^{-1}$$

where $a^{(j)} \in KF$, $l(a^{(j)}) < l$. Therefore

$$l(\partial_{ij} u) < l \quad \text{for} \quad i \neq j. \tag{8}$$

Further, $(\partial_j u) x_j = a^{(j)} x_j - w^{(j)}$ and so

$$l(\partial_j ((\partial_j u) x_j)) < l. \tag{9}$$

Since $u \in \Delta^n$, it follows from Proposition 0.5.4 that $\partial_j u \in \Delta^{n-1}$ and $\partial_{ij} u \in \Delta^{n-2}$ for every i and j. The first inclusion implies that $(\partial_j u) x_j \in \Delta^{n-1}$ and so, again by 0.5.4, $\partial_j ((\partial_j u) x_j) \in \Delta^{n-2}$. By induction hypothesis,

$$l(\partial_{ij} u) \geq \frac{n-2}{2} \quad \text{and} \quad l(\partial_j ((\partial_j u) x_j)) \geq \frac{n-2}{2} \tag{10}$$

(or the corresponding elements are zeroes!). It follows from (8) and (10) that $l - 1 \geq (n/2) - 1$, that is, $l \geq n/2$, which contradicts the assumption.

The same contradiction follows from (9) and (10). Thus it is mandatory that

$$\partial_{ij}u = 0 \quad \text{for} \quad i \neq j; \qquad \partial_j((\partial_j u)x_j) = 0. \tag{11}$$

The last equality can be rewritten as $0 = \partial_j((\partial_j u)x_j) = \partial_{jj}u + \partial_j u$, whence

$$\partial_{jj}u = -\partial_j u. \tag{12}$$

Now, using 0.5.2, 0.5.4, (11) and (12), for arbitrary j we have

$$\partial_j u = \partial_j u(1) + \sum_i \partial_{ij}u \cdot (x_i - 1) = -\partial_j u \cdot (x_j - 1)$$

and so $\partial_j u = 0$. Consequently,

$$u = u(1) + \sum_j \partial_j u \cdot (x_j - 1) = 0. \quad \square$$

0.5.6. Corollary (Magnus [57]). *For arbitrary K the augmentation ideal of KF is residually nilpotent, that is, $\bigcap_{n=1}^{\infty} \Delta^n = 0$.* $\square$

0.5.7. Corollary. *If $u(1) = v(1)$, $\partial_i u(1) = \partial_i v(1)$, $\partial_{ij}u(1) = \partial_{ij}v(1)$, ..., then $u = v$.*

Proof. If $w = u - v$, then $w(1) = \partial_i w(1) = \cdots = 0$. By Proposition 0.5.4, $w \in \bigcap \Delta^n = 0$, so that $u = v$. $\square$

The Magnus embedding. For arbitrary $u \in KF$, the *formal decomposition of u in the Taylor series* is the expression

$$u = u(1) + \sum(\partial_i u(1))(x_i - 1) + \sum(\partial_{ij}u(1))(x_i - 1)(x_j - 1) + \ldots$$
$$+ \sum(\partial_{i_1 \ldots i_n}u(1))(x_{i_1} - 1)(x_{i_n} - 1) + \ldots . \tag{13}$$

In particular, one can easily calculate the formal Taylor series for x_i and x_i^{-1}:

$$x_i = 1 + (x_i - 1),$$

$$x_i^{-1} = 1 - (x_i - 1) + (x_i - 1)^2 - (x_i - 1)^3 + \ldots . \tag{14}$$

By 0.5.7, every $u \in KF$ is uniquely determined by its Taylor series. This observation has important consequences. Namely, let $K\langle\langle Z \rangle\rangle = K\langle\langle z_1, z_2, \dots \rangle\rangle$ be the K-algebra of formal power series in countably many noncommutative indeterminates $z_1, z_2, \dots$. Each element of $K\langle\langle Z \rangle\rangle$ is a series of the form

$$w = w_{(0)} + w_{(1)} + \cdots + w_{(n)} + \cdots$$

where $w_{(n)}$, the *homogeneous component of degree* n, is a finite sum of monomials $\lambda z_{i_1} z_{i_2} \dots z_{i_n}$ ($\lambda \in K$). Now, using (13), define a map $\mu : KF \to K\langle\langle Z \rangle\rangle$ as follows:

$$u^\mu = u(1) + \sum (\partial_i u(1)) z_i + \sum (\partial_{ij} u(1)) z_i z_j + \cdots \tag{15}$$
$$+ \sum (\partial_{i_1 \dots i_n} u(1)) z_{i_1} \dots z_{i_n} + \cdots .$$

Corollary 0.5.7 guarantees that μ is injective. Moreover, a straightforward verification shows that μ is a homomorphism of K-algebras. Taking this into account, from now on we will identify KF with its image $(KF)^\mu$ in $K\langle\langle Z \rangle\rangle$. In particular, we can now write

$$x_i = 1 + z_i \quad \text{or} \quad z_i = x_i - 1,$$

but the formula (14) acquires the following shape:

$$x_i^{-1} = 1 - z_i + z_i^2 - z_i^3 + \cdots . \tag{16}$$

In other words, if we denote by x_i the element $1 + z_i$ of $K\langle\langle Z \rangle\rangle$, the K-subalgebra of $K\langle\langle Z \rangle\rangle$ generated by all the x_i and x_i^{-1} is isomorphic to KF, x_i's being free generators of F.

Note. The above embedding of KF into the algebra of formal power series was in essence discovered by Magnus [57]. It should not be confused with another Magnus embedding giving a faithful representation of certain abelian extensions F/R' by 2×2 matrices of a special form [59].

Denote by Q the set of all series without a constant term, that is, the set of all

$$u = u_{(0)} + u_{(1)} + \cdots + u_{(n)} + \dots \in K\langle\langle Z \rangle\rangle$$

such that $u_{(0)} = 0$. Clearly Q is an ideal of $K\langle\langle Z\rangle\rangle$, and so is Q^n for each n. Since KF is a subalgebra of $K\langle\langle Z\rangle\rangle$, the intersection $Q^n \cap KF$ is an ideal of KF.

0.5.8. Proposition. $Q^n \cap KF = \Delta^n$.

Proof. If $u \in \Delta$, then $u(1) = 0$, and it follows from (15) that $\Delta \subseteq Q$. Therefore $\Delta^n \subseteq Q^n \cap KF$. Conversely, let $u \in Q^n \cap KF$. By (7),

$$u = u_{(1)} + u_{(2)} + \cdots + u_{(n-1)} + u^{(n)}$$

where $u_{(i)}$ $(i < n)$ is a homogeneous element of degree i and

$$u^{(n)} = \sum (\partial_{i_1 \dots i_n}) z_{i_1} \dots z_{i_n} \in \Delta^n.$$

Therefore $u_{(1)} + \cdots + u_{(n-1)} = u - u^{(n)} \in Q^n$, whence $u_{(1)} = \cdots = u_{(n-1)} = 0$. Thus $u = u^{(n)} \in \Delta^n$. $\square$

0.5.9. Corollary. *For arbitrary* n
(i) Δ^n *is a free (right) KF-module;*
(ii) Δ^n/Δ^{n+1} *is a free K-module.*

Proof. It follows from the above that all the possible monomials $z_{i_1} \dots z_{i_n}$ of degree n form a KF-basis for Δ^n and a K-basis for Δ^n over Δ^{n+1}. $\square$

TWO TECHNICAL RESULTS. Concluding this section, we will prove two assertions similar to well known group-theoretic results of Higman [36] and Powell [82] (see also [68, 33.37 and 33.43]).

Let $G = \prod^* G_i$ be the free product of an at most countable family of groups $G_1, G_2, \dots$. Denote by θ_i the endomorphism of G taking G_i to 1 and acting identically on each G_j with $j \neq i$. It can be naturally extended to an endomorphism of the regular representation $\text{Reg } G = (KG, G)$ which will be denoted by the same symbol θ_i. An element $0 \neq m \in KG$ is said to be a *monomial of degree* n, if $m = (g_1 - 1)(g_2 - 1) \dots (g_n - 1)$ where $1 \neq g_i \in G_{\lambda(i)}$. The set $\text{Supp } m = \{\lambda(1), \dots, \lambda(n)\}$ is called the *support* of m.

Take the endomorphism θ_i of (KG, G) and let A_i be its kernel in KG. Then A_i is contained in the augmentation ideal Δ of KG. For a finite nonempty set I of natural numbers, denote $A_I = \bigcap_{i \in I} A_i$. Furthermore, let $A_\emptyset = \Delta$.

0.5.10. Lemma. *Any element from A_I is a K-linear combination of monomials whose supports contain I.*

P r o o f. We proceed by induction on $|I|$. For $I = \emptyset$ there is nothing to prove. Suppose the lemma has been already proved for $|I| = n - 1$, and let $I = \{i_1, \ldots, i_{n-1}, i\}$. If $u \in A_I$, then $u \in A_{I'}$ where $I' = \{i_1, \ldots, i_{n-1}\}$. By induction hypothesis, $u = \sum_{j=1}^{s} \alpha_j m_j$, where $\alpha_j \in K$ and m_j are monomials such that $I' \subseteq \operatorname{Supp} m_j$ for every j. Without loss of generality, we may assume that for some r

$$
\begin{aligned}
i \in \operatorname{Supp} m_j && \text{if} && j = 1, \ldots, r; \\
i \notin \operatorname{Supp} m_j && \text{if} && j = r + 1, \ldots, s.
\end{aligned}
$$

Apply now θ_i to the equality

$$ u = \sum_{i}^{r} \alpha_j m_j + \sum_{r+1}^{s} \alpha_j m_j. $$

Since $u\theta_i = 0$, and since for any monomial m we have $m\theta_i = 0$ if $i \in \operatorname{Supp} m$, and $m\theta_i = m$ otherwise, it follows that $\sum_{r+1}^{s} \alpha_j m_j = 0$. Therefore $u = \sum_{1}^{r} \alpha_j m_j$ and, since $i \in \operatorname{Supp} m_j$ for every $j = 1, \ldots, r$, the proof is completed. $\square$

Let $w \in KG$. Since $\theta_i^2 = \theta_i$, we have

$$ w(1 - \theta_i)\theta_i = 0, \quad \text{that is,} \quad w(1 - \theta_i) \in A_i. \tag{17} $$

Evidently, every A_i is invariant under every θ_j. Therefore it follows from (17) that $w(1 - \theta_i)(1 - \theta_j) \in A_i \cap A_j$, whence by induction

$$ w(1 - \theta_{i_1})(1 - \theta_{i_2}) \ldots (1 - \theta_{i_n}) \in A_{\{i_1, i_2, \ldots, i_n\}}. \tag{18} $$

On the other hand, expanding the left side of this formula, we obtain

$$w(1 - \theta_{i_1})(1 - \theta_{i_2}) \ldots (1 - \theta_{i_n}) = w + \sum (-1)^r w \theta_{j_1} \ldots \theta_{j_r} \qquad (19)$$

where the sum is taken over all subsequences $j_1, \ldots, j_r$ of the sequence $i_1, \ldots, i_n$. Thus, combining (18) and (19), we have:

0.5.11. Lemma. *Let $w = w(x_1, \ldots, x_n) \in KG$, where $G = \prod^* G_i$. Then for arbitrary n the element w can be written in the form*

$$w = u + v_1 + \cdots + v_t$$

where $u \in A_{\{1,2,\ldots,n\}}$ and each v_i has the form $v_i = \pm w \theta_{j_1} \ldots \theta_{j_r}$ ($r \geq 1, 1 \leq j_1 < \cdots < j_r \leq n$). $\square$

In the particular case $G = F$, Lemmas 0.5.10 and 0.5.11 imply:

0.5.12. Corollary. *If $w \in KF$, then for arbitrary n*

$$w = u + v_1 + \cdots + v_t$$

where $u \in \Delta^n$ and each v_i has the form $v_i = \pm w \theta_{j_1} \ldots \theta_{j_r}$ ($r \geq 1, 1 \leq j_1 < \cdots < j_r \leq n$). $\square$

It is not hard to see that the last assertion can be also deduced from Proposition 0.5.3.

Chapter 1

STABLE VARIETIES AND HOMOGENEITY

Recall that a variety of group representations is said to be *stable of class n* (n-stable) if it satisfies the identity

$$y \circ (x_1 - 1)(x_2 - 1)\ldots(x_n - 1) \equiv 0$$

or, in other words, if it is a subvariety of $\mathcal{S}^n$. In the present section we are concerned with some aspects of the theory of stable varieties.

Section 1.1 deals with "finiteness properties" of such varieties. In particular, it is proved that every stable variety has a finite basis of identities (provided the ground ring is noetherian), a finite axiomatic rank and a finite basis rank.

In the remaining part of the chapter the ground ring K is a field (although this restriction sometimes is superfluous). In §1.2–1.5 our considerations are essentially based on the idea of homogeneity. First of all, the Magnus embedding of KF into the algebra of formal power series makes it possible to introduce, in a natural manner, the notion of a homogeneous variety of group representations. This notion leads to a number of interesting questions; furthermore, it turns out to be useful in studying stable varieties. One of the main results of the chapter (Theorem 1.3.1) states that if K is a field of characteristic zero, then there exists a canonical one-to-one correspondence between the so-called homogeneous Magnus varieties of group representations over K and *all* varieties of (associative) algebras over K. This theorem enables us to apply the deeply elaborated theory of varieties of algebras to the study of varieties of group representations. In particular, we will see that there exists a canonical bijection between *homogeneous n-stable* varieties of representations and *all* n-nilpotent varieties of algebras.

One application of this observation is given in § 1.4 where, following a paper of Grinberg [23], we describe all 4-stable varieties of representations over a field of characteristic zero.

On the other hand, non-homogeneous stable varieties do exist (Theorem 1.5.1), and so (fortunately!) stable varieties of group representations can not be reduced to nilpotent varieties of algebras.

In § 1.6, using a recent theorem of Zel'manov [105] on the nilpotency of Lie algebras with the Engel condition, we prove that every unipotent representation over a field of characteristic zero is stable. In other words, in characteristic zero the identity $(x - 1)^n$ implies the identity $(x_1 - 1)(x_2 - 1) \ldots (x_N - 1)$ for some $N = N(n)$. This result will be used essentially in the next chapter.

1.1. Finiteness properties

We begin with one simple but useful assertion which is valid over an arbitrary ground ring K.

1.1.1. Proposition. *The identities of any stable variety of class n follow from $(x_1 - 1) \ldots (x_n - 1)$ and identities in $n - 1$ variables.*

P r o o f. Let $\mathcal{X}$ be an n-stable variety and $I = \mathrm{Id}\,\mathcal{X}$. By 0.5.3, an arbitrary $u \in I$ can be written in the form

$$u = \lambda + \sum \lambda_i z_i + \sum \lambda_{ij} z_i z_j + \cdots + \sum \lambda_{i_1 \ldots i_{n-1}} z_{i_1} \ldots z_{i_{n-1}} + w$$

where $\lambda, \lambda_i, \lambda_{ij}, \cdots \in K$, $z_i = x_i - 1$, $w \in \Delta^n$ and all sums are finite. Since $\Delta^n \subseteq I$, we have $u - w \in I$. Here $u - w$ is a polynomial of degree at most $n - 1$ in variables z_i and can be uniquely presented as a sum of normal polynomials (a polynomial is *normal* if all its monomials involve just the same set of variables; for example, $\alpha z_1 + \beta z_1^2 + \gamma z_1^5$ and $\alpha z_1 z_2 + \beta z_2 z_1 z_2 + \gamma z_1 z_2^4$ are normal polynomials). Using deletions of z_i's (that is, the endomorphisms of F taking the corresponding x_i's to 1), we see that all these normal polynomials also belong to I. But a normal polynomial of degree at most $n - 1$ cannot involve more than $n - 1$ variables, and it remains to notice that w is a consequence of $z_1 z_2 \ldots z_n$. $\square$

1.1.2. Corollary. *The axiomatic rank of an n-stable variety does not exceed n.* $\square$

Of course, there exist stable varieties whose axiomatic rank is strictly less than the class of stability. For instance, consider the variety $\mathcal{U}_2 = [(x - 1)^2]$. Then $r_a(\mathcal{U}_2) = 1$, but in § 1.4 we will prove that $\mathcal{U}_2$ is a stable variety of class 3 (provided $1/2 \in K$).

1.1.3. Corollary. *Let the ground ring K be noetherian. Then every stable variety of group representations over K is finitely based.*

Proof. By Proposition 1.1.1, every n-stable verbal ideal of KF is generated, as a completely invariant ideal, by $z_1 z_2 \ldots z_n$ and a set $\{u_k\}$, where each u_k is a polynomial of degree $\leq n - 1$ involving $\leq n - 1$ variables z_i. Up to permutations of z_i, we may assume that each u_k is a polynomial in the variables $z_1, \ldots, z_{n-1}$. Then $\{u_k\}$ is contained in the K-submodule M of KF generated by the monomials of degree $\leq n - 1$ in the indeterminates $z_1, \ldots, z_{n-1}$. Since M is a finitely generated module over a noetherian ring, it is a noetherian module. Therefore the K-submodule generated by $\{u_k\}$ is generated by some finite subset. This subset together with $z_1 z_2 \ldots z_n$ constitutes a finite verbal basis for I. $\square$

Corollary 1.1.3 can be generalized in several directions. We will prove one of such generalizations which was inspired by a theorem of Higman [36] on varieties of groups.

1.1.4. Theorem (Grinberg [22]). *Let K be a noetherian ring. If $\mathcal{X}$ is a stable variety of group representations over K and Θ is a finitely based variety of groups, then the variety $\mathcal{X} \times \Theta$ is finitely based.*

The proof is similar to that of Higman's theorem (cf. [36] or [68, 34.23]). First we establish one technical fact. Denote by τ_n the endomorphism of the group F defined by the rule $x_i \tau_n = x_{i+n}$ $(i = 1, 2, \ldots)$.

1.1.5. Lemma. *If* $\mathcal{X}$ *is determined by an identity* $u = u(x_1, \ldots, x_m) \in KF$ *and* Θ *is determined by an identity* $f = f(x_1, \ldots, x_n)$, *then* $\mathcal{X} \times \Theta$ *is determined by the set of identities*

$$u(f_1^{\pm 1} \ldots f_k^{\pm 1}, \ f_{k+1}^{\pm 1} \ldots f_{2k}^{\pm 1}, \ \ldots, \ f_{(m-1)k+1}^{\pm 1} \ldots, f_{mk}^{\pm 1}) \qquad (1)$$

where $k = 1, 2, \ldots$ *and* $f_i = f\tau_n^{i-1}$.

P r o o f. From the definition of $\mathcal{X} \times \Theta$ it is easy to understand that if $I = \mathrm{Id}\,\mathcal{X}$ and $H = \Theta^*(F)$ is the Θ-verbal subgroup of F, then $\mathrm{Id}(\mathcal{X} \times \Theta)$ consists of all elements of the form

$$w(h_1, \ldots, h_s), \quad \text{where} \quad w(x_1, \ldots, x_s) \in I, \ h_i \in H.$$

It follows that $\mathrm{Id}(\mathcal{X} \times \Theta)$ is generated, as a fully invariant ideal, by the set of identities

$$u(h_1, \ldots, h_m), \quad \text{where} \quad h_i \in H.$$

Every word from H can be presented in the form $(f_1^{\pm 1} f_2^{\pm 1} \ldots f_r^{\pm 1})^\alpha$, where $\alpha \in \mathrm{End}\,F$ and the f_i are copies of f written in pairwise disjoint sets of the variables x_j. Therefore it is clear enough that the system of identities (1) is a verbal basis for $\mathrm{Id}(\mathcal{X} \times \Theta)$. $\square$

Let s be a positive integer. For every sequence of natural numbers $m_1 < m_2 < \cdots < m_t$, where $t \leq s - 1$, define an endomorphism $\pi(m_1, \ldots, m_t)$ of F by the rule

$$x_i \pi(m_1, \ldots, m_t) = \begin{cases} x_j & \text{if } i = m_j, \\ 1 & \text{otherwise.} \end{cases}$$

In addition, we set $x_i \pi(0) = 1$ for each i.

1.1.6. Lemma. *Every word* $w \in KF$ *is equivalent to a set consisting of all words of the form* $w\pi(m_1, \ldots, m_t)$ *with* $t < s$, *plus a word from* Δ^s.

P r o o f. Apply 0.5.12 to w, then

$$w = u + v_1 + v_2 + \cdots + v_q$$

where $u \in \Delta^s$, but each v_i has the form $v_i = \pm w\theta_{j_1} \ldots \theta_{j_r}$ and involves strictly fewer variables than w. Since all the v_i are consequences of w, the word u is a consequence of w as well. Therefore w is equivalent to the set $\{u, v_1, \ldots, v_q\}$. If some v_i involves more than s variables, we can apply the same argument to this v_i and replace it by a word from Δ^s plus a finite set of words each involving fewer variables than v_i. Continuing this process, we eventually come to a finite set of words which either involve at most $s - 1$ variables or belong to Δ^s. The latter can be replaced by a single word from Δ^s, while the former can be written in the variables $x_1, \ldots, x_{s-1}$. Taking into account the procedure of obtaining these words in $x_1, \ldots, x_{s-1}$, we see that they all are of the desired form $w\pi(m_1, \ldots, m_t)$. $\square$

Proof of Theorem 1.1.4. Since every finite set of identities is equivalent to a single identity, we may assume that $\mathcal{X}$ is determined by an identity $u = u(x_1, \ldots, x_m) \in KF$ (in view of 1.1.3) and Θ is determined by an identity $f = f(x_1, \ldots, x_n) \in F$. By Lemma 1.1.5, $\mathcal{X} \times \Theta$ is determined by the set (1). Denote $\mathrm{Id}(\mathcal{X} \times \Theta) = I$. Since $\mathcal{X}$ is stable (say, of class s), for any $g_1, \ldots, g_s \in \Theta^*(F)$ we have

$$(g_1 - 1) \ldots (g_s - 1) \in \mathrm{Id}(\mathcal{X} \times \Theta).$$

In particular, $(f_1 - 1) \ldots (f_s - 1) \in I$, where the group words $f_i = f\tau_n^{i-1}$ are defined as in 1.1.5.

Consider a free group Φ of countable rank with free generators $y_1, y_2, \ldots$ and let M be the subset of $K\Phi$ consisting of

(i) all words $u(y_1^{\pm 1} \ldots y_k^{\pm 1}, \ y_{k+1}^{\pm 1} \ldots y_{2k}^{\pm 1}, \ \ldots, \ y_{(m-1)k+1}^{\pm 1} \ldots y_{mk}^{\pm 1})$, where $k = 1, 2, \ldots$, and

(ii) the word $(y_1 - 1)(y_2 - 1) \ldots (y_s - 1)$.

Apply Lemma 1.1.6 to this set M. Denote now by $\pi(m_1, \ldots, m_t)$, where $m_1 < \cdots < m_t$ and $t \le s - 1$, the endomorphism of Φ taking y_i to y_j if $i = m_j$, and to 1 otherwise. By 1.1.6, every word w from (i) is equivalent to a set consisting of all the $w\pi(m_1, \ldots, m_t)$ and an element from Δ^s. Note that the latter is a consequence of the word $(y_1 - 1) \ldots (y_s - 1)$. Since each element from $(K\Phi)^{\pi(m_1, \ldots, m_t)}$ involves at most $s - 1$ elements y_i, it is enough to take the words (i) only with $k = s - 1$. In other words, M is equivalent in $K\Phi$ to its subset M^* consisting of the words

$$u\big(y_1^{\pm 1} \ldots y_{s-1}^{\pm 1}, \ y_s^{\pm 1} \ldots y_{2(s-1)}^{\pm 1}, \ \ldots, \ y_{(m-1)(s-1)+1}^{\pm 1} \ldots y_{m(s-1)}^{\pm 1}\big)$$

and $(y_1 - 1)\ldots(y_s - 1)$. Note that M^* is finite.

Let now $\eta : \Phi \to F$ be the homomorphism defined by the rule $y_i^\eta = f_i$. Lemma 1.1.5 guarantees that the set M^η is a verbal basis for I (that is, I is generated by M^η as a verbal ideal). If we establish that M^η is equivalent in KF to its finite subset $M^{*\eta}$, the proof will be completed.

Since M and M^* are equivalent in $K\Phi$, it follows that M can be obtained from M^* by means of addition, multiplication by arbitrary elements from $K\Phi$, and endomorphisms of Φ. The first two operations are preserved under η. However, an arbitrary endomorphism of Φ *does not correspond*, in general, to any endomorphism of F. Fortunately, in proving that M and M^* are equivalent in $K\Phi$, we used only the endomorphisms $\pi(m_1, \ldots, m_t)$. But it is not hard to understand that these endomorphisms *do correspond* to certain endomorphisms of F which map blocks of variables $x_{(i-1)n+1}, \ldots, x_{in}$ either onto other blocks of this form, or onto 1 (it is essential that the sets of variables ocurring in the words $f_i = y_i^\eta$ are pairwise non-overlapping). This eventually shows that the sets M^η and $M^{*\eta}$ are equivalent in KF, as required. $\square$

Note. If $\mathcal{X}$ is an arbitrary finitely based variety of representations and Θ a finitely based variety of groups, then $\mathcal{X} \times \Theta$ need not be finitely based. Indeed, consider the Burnside varieties of groups $\mathfrak{B}_2$ and $\mathfrak{B}_4$ (determined by $x^2 \equiv 1$ and $x^4 \equiv 1$ respectively). Then

$$(\omega\mathfrak{B}_4) \times \mathfrak{B}_2 = \omega(\mathfrak{B}_4\mathfrak{B}_2)$$

where $\omega\mathfrak{B}_4$ and $\mathfrak{B}_2$ are finitely based. But the group variety $\mathfrak{B}_4\mathfrak{B}_2$ is not finitely based (Kleiman [41]) and so, by 0.3.3, $\omega(\mathfrak{B}_4\mathfrak{B}_2)$ is not finitely based either.

We turn now to the basis rank of stable varieties. The following statement is analogous to Theorem 35.11 from [68].

1.1.7. Proposition. *The basis rank of an n-stable variety of group representations does not exceed n.*

P r o o f. Let $\mathcal{X}$ be an n-stable variety and let $\mathrm{Fr}_n\mathcal{X} = (A, G)$ be the free cyclic representation of rank n in $\mathcal{X}$. We prove that for each $r > n$

the representation $\mathrm{Fr}_r\mathcal{X}$ can be isomorphically embedded into the direct product of $\binom{r}{n}$ copies of $\mathrm{Fr}_n\mathcal{X}$. This will imply that $\mathcal{X} = \mathrm{var}(\mathrm{Fr}_n\mathcal{X})$, so that $r_b(\mathcal{X}) \leq n$.

Let $\{a;\ g_1,\ldots,g_n\}$ (where $a \in A$, $g_1,\ldots,g_n \in G$) be a system of free generators of $\mathrm{Fr}_n\mathcal{X}$. Consider the set M consisting of all possible n-tuples $m = (m_1,\ldots,m_n)$, where $1 \leq m_1 < \cdots < m_n \leq r$, and let

$$(\mathrm{Fr}_n\mathcal{X})^M = (A^M, G^M) = (\bar{A}, \bar{G})$$

be the direct M-power of $\mathrm{Fr}_n\mathcal{X}$. Note that $|M| = \binom{r}{n}$. Choose in $\bar{G}$ the elements $\bar{g}_1,\ldots,\bar{g}_r$ defined as follows:

$$\bar{g}_i(m) = \begin{cases} g_j & \text{if } i = m_j; \\ 1 & \text{otherwise.} \end{cases} \tag{2}$$

Furthermore, let $\bar{a}$ be the element of $\bar{A}$ such that $\bar{a}(m) = a$ for each $m \in M$. Denote by $\rho = (B, H)$ the subrepresentation of $(\bar{A}, \bar{G})$ generated by the set $\{\bar{a};\ \bar{g}_1,\ldots,\bar{g}_r\}$, that is, $H = \langle \bar{g}_1,\ldots,\bar{g}_r \rangle$ and $B = \bar{a} \circ KH$. We prove that $\rho \simeq \mathrm{Fr}_r\mathcal{X}$.

It suffices to show that if for some $u = u(x_1,\ldots,x_r) \in KF$ the equality

$$\bar{a} \circ u(\bar{g}_1,\ldots,\bar{g}_r) = 0 \tag{3}$$

holds in ρ, then $u(x_1,\ldots,x_r)$ is an identity of $\mathcal{X}$. First of all we note that (3) must be satisfied componentwise, that is, for every $m = (m_1,\ldots,m_n) \in M$ the equality

$$a \circ u(\bar{g}_1(m),\ldots,\bar{g}_r(m)) = 0 \tag{4}$$

holds in $\mathrm{Fr}_n\mathcal{X}$. Now we again use the endomorphisms $\pi(m_1,\ldots,m_n)$ of F. Recall that

$$x_i\pi(m_1,\ldots,m_n) = \begin{cases} x_j & \text{if } i = m_j; \\ 1 & \text{otherwise.} \end{cases} \tag{5}$$

One can see from (2) and (5) that $u(\bar{g}_1(m),\ldots,\bar{g}_r(m)) \in KG$ is obtained from $u\pi(m_1,\ldots,m_n) \in KF$ by substituting g_j's for the corresponding x_j's. Hence (4) can be rewritten as

$$a \circ [(u\pi(m_1,\ldots,m_n))(g_1,\ldots,g_n)] = 0.$$

This is a relation between the free generators of $\mathrm{Fr}_n \mathcal{X}$, therefore $u\pi(m_1, \ldots$ $\ldots, m_n)$ is an identity of $\mathrm{Fr}_n \mathcal{X}$. Since $u\pi(m_1, \ldots, m_n)$ involves just n variables, it is an identity of the whole variety $\mathcal{X}$. This implies, of course, that every word $u\pi(m_1, \ldots, m_t)$, where $m_1 < \cdots < m_t$ and $t \leq n-1$, is an identity of $\mathcal{X}$.

By Lemma 1.1.6, u is equivalent to the set consisting of all possible words $u\pi(m_1, \ldots, m_t)$, where $m_1 < \cdots < m_t$ and $t \leq n-1$, and an element from Δ^n. Since all these words are identities of $\mathcal{X}$, so is u. $\square$

1.2. Homogeneous varieties

It is well known that every variety $\mathcal{M}$ of linear algebras, say over a field of characteristic zero, possesses an important property of homogeneity: for each identity of $\mathcal{M}$, all its homogeneous components are also identities of this variety (Mal'cev [61]). As a result, one can reduce arbitrary identities of algebras to their homogeneous components which, in turn, can be reduced to multilinear identities. This fact is of principal importance for the theory of varieties of algebras because multilinear identities are much more accessible to study and classification than arbitrary ones.

In the theory of varieties of group representations the situation is more complicated. Indeed, identities of group representations are elements of the free group algebra KF, and it is clear that from the usual presentation

$$u = u(x_1, \ldots, x_n) = \sum \lambda_i f_i(x_1, \ldots, x_n) \quad (\lambda_i \in K, f_i(x_1, \ldots, x_n) \in F)$$

of an arbitrary $u \in KF$ one cannot extract any "homogeneous components" of u. However, elements of the algebra KF have another canonical presentation, based on the Magnus embedding (see § 0.5). Namely, KF is identified with its natural image in the algebra $K\langle\langle Z \rangle\rangle$ of formal power series in a countable set of variables $Z = \{z_1, z_2, \ldots\}$. Every $u \in KF$ is uniquely presented as a formal power series

$$u = u_{(0)} + u_{(1)} + \cdots + u_{(n)} + \cdots$$

where $u_{(n)} = \sum \lambda_{i_1 \ldots i_n} z_{i_1} \ldots z_{i_n}$ is a finite K-linear combination of monomials of degree n in variables $z_1, z_2, \ldots$ (in particular, $x_i = 1 + z_i$). The element $u_{(n)}$ is called the *n-th homogeneous component* of the given u. This naturally leads to the following definition [24]: a variety of group representations $\mathcal{X}$ is called *homogeneous* if for any identity u of $\mathcal{X}$ all homogeneous components $u_{(n)}$ are identities of $\mathcal{X}$ as well.

Contrary to varieties of algebras, a variety of group representations need not be homogeneous (even over a field of characteristic zero). For example, the variety $\omega\mathcal{B}_2 = [x^2 - 1]$ is not homogeneous, for

$$x_1^2 - 1 = (x_1 - 1)^2 + 2(x_1 - 1) = z_1^2 + 2z_1,$$

but $2z_1$, evidently, is not an identity of $\omega\mathcal{B}_2$. Later we will present more interesting examples.

In general, one can say that the requirement of homogeneity is rather strong; the fact that a variety is homogeneous gives essential information about this variety. In particular, the idea of homogeneity plays an important role in the investigation of stable varieties.

In the present section several useful properties of homogeneous varieties will be established. The first three assertions were proved in [98]. For convenience, elements from KF will sometimes be called *words* (as opposed to arbitrary series from $K\langle\langle Z \rangle\rangle$). Let $u \in KF$ be an arbitrary word and let $u = u_{(0)} + u_{(1)} + \cdots + u_{(n)} + \ldots$ be its decomposition as a formal power series in the variables $z_i = x_i - 1$. The word $u_{(n)}$ is homogeneous of degree n. It is clear that $u_{(n)}$ can be uniquely decomposed as a sum of multihomogeneous words, that is, words which are homogeneous in each z_i. They are called the *multihomogeneous components* of the given u.

It is well known that every variety of linear algebras over an infinite field is *multihomogeneous*, i.e. all multihomogeneous components of its identities are identities of this variety as well. On the other hand, it has just been shown that a variety of group representations need not even be homogeneous. However:

1.2.1. Lemma. *If a variety of group representations over a field of characteristic zero is homogeneous, then it is multihomogeneous.*

P r o o f. We use a standard argument. Let $\mathcal{X}$ be a homogeneous variety of representations. This means that the verbal ideal $\operatorname{Id}\mathcal{X} = I$ is homogeneous, that is, for each $u \in I$ all the $u_{(n)}$ belong to I as well. It suffices to show that if $u \in I$ is a homogeneous word of some degree m, then each multihomogeneous component of u belongs to I. Let $z_1, \ldots, z_n$ be the variables involved in u; then u can be presented in the form $u = u_0 + u_1 + \cdots + u_s$, where u_i is homogeneous of degree i in z_1. Denoting equality modulo I by $\equiv$, we can rewrite this as

$$u_0 + u_1 + \cdots + u_s \equiv 0. \tag{1}$$

Consider the endomorphism $\varphi : x_1 \mapsto x_1^2$ of F (that is, $x_1^\varphi = x_1^2$ and $x_i^\varphi = x_i$ for $i \neq 1$). Then $z_1^\varphi = 2z_1 + z_1^2$ and hence, applying φ to (1), we obtain

$$u_0 + 2u_1 + 2^2 u_2 + \cdots + 2^s u_s + \quad \text{terms of degree} > m \quad \equiv 0.$$

Since I is a homogeneous ideal, it follows that

$$u_0 + 2u_1 + 2^2 u_2 + \cdots + 2^s u_s \equiv 0.$$

Next, apply to (1) the endomorphism $x_1 \mapsto x_1^3$. Then $z_1 \mapsto 3z_1 +$ higher degree terms, and we obtain

$$u_0 + 3u_1 + 3^2 u_2 + \cdots + 3^s u_s \equiv 0.$$

Continuing this process, we eventually come to the following systems of equalities modulo I:

$$u_0 + u_1 + u_2 + \cdots + u_s \equiv 0,$$
$$u_0 + 2u_1 + 2^2 u_2 + \cdots + 2^s u_s \equiv 0,$$
$$u_0 + 3u_1 + 3^2 u_2 + \cdots + 3^s u_s \equiv 0,$$
$$\cdots \cdots \cdots \cdots \cdots \cdots \cdots \cdots \cdots$$
$$u_0 + (s+1)u_1 + (s+1)^2 u_2 + \cdots + (s+1)^s u_s \equiv 0.$$

Since char $K = 0$, the determinant of this system is different from zero. Hence $u_0 \equiv u_1 \equiv \cdots \equiv u_s \equiv 0$. We can now apply the same argument to the words u_i and the variable z_2, and so on. $\square$

Note. In ring theory the above argument is valid over an arbitrary infinite field K. In our case it is not true, for we can use not all possible endomorphisms of the algebra KF, but only those induced by endomorphisms of the group F. In informal language this means that instead of z_i we can substitute not all possible λz_i ($\lambda \in K$), but only $n z_i$ ($n \in \mathbb{Z}$). Therefore the condition char $K = 0$ is essential. However, it is easy to see that if K is any commutative ring, the proof of Lemma 1.2.1 remains valid for such u whose degree m satisfies the condition $1/(m-1)! \in K$.

Let $K\langle Z\rangle$ be the absolutely free associative algebra (without 1!) over K on a countable set of variables $Z = \{z_1, z_2, \dots\}$. It is naturally contained in the algebra $K\langle\langle Z\rangle\rangle$ and, moreover, in the subalgebra KF of $K\langle\langle Z\rangle\rangle$. Let $u = u(z_1, \dots, z_n)$ be a multihomogeneous element from $K\langle Z\rangle$, $m_i = \deg_{z_i} u$ and $m = m_1 + \cdots + m_n$. Denote by $\lim u = (\lim u)(z_1, \dots, z_m)$ the *full linearization of* u, that is, the multilinear part of the polynomial

$$u^* = u(z_1 + \cdots + z_{m_1},\ z_{m_1+1} + \cdots + z_{m_1+m_2},\ \dots,\ z_{m_1+\cdots+m_{n-1}+1} + \cdots + z_m).$$

Example. Let $u(z_1, z_2) = z_1^3 z_2^2$. Then $m_1 = 3, m_2 = 2$ and $u^* = (z_1 + z_2 + z_3)^3 (z_4 + z_5)^2$, whence

$$\lim u = z_1 z_2 z_3 z_4 z_5 + z_1 z_2 z_3 z_5 z_4 + z_2 z_1 z_3 z_4 z_5 + \cdots$$

$$= \sum_{\sigma, \tau} z_{\sigma(1)} z_{\sigma(2)} z_{\sigma(3)} z_{\tau(4)} z_{\tau(5)}$$

where σ is an arbitrary permutation of $\{1, 2, 3\}$ and τ an arbitrary permutation of $\{4, 5\}$.

1.2.2. Lemma. *If* char $K = 0$, *then for any homogeneous verbal ideal I of KF*

$$u \in I \iff \lim u \in I.$$

Proof. We use the same notation as above. Consider an endomorphism φ of F such that

$$x_1^\varphi = x_1 \dots x_{m_1},\ x_2^\varphi = x_{m_1+1} \dots x_{m_1+m_2},\ \dots,\ x_n^\varphi = x_{m_1+\cdots+m_{n-1}+1} \dots x_m.$$

Then

$$z_1^\varphi = z_1 + z_2 + \cdots + z_{m_1} + \text{ higher degree terms},$$

$$z_2^\varphi = z_{m_1+1} + \cdots + z_{m_1+m_2} + \text{ higher degree terms},$$

$$\dots \dots \dots \dots \dots \dots \dots \dots \dots$$

$$z_n^\varphi = z_{m_1+\cdots+m_{n-1}+1} + \cdots + z_m + \text{ higher degree terms}.$$

Suppose that $u \in I$. Then

$$u^{\varphi} = u(z_1 + \cdots + z_{m_1} + \cdots, \ \ldots, \ z_{m_1 + \cdots + m_{n-1} + 1} + \cdots + z_m + \ldots) \in I.$$

It is evident that the multilinear with respect to $z_1, \ldots, z_m$ component of u^{φ} is exactly the polynomial $\operatorname{lin} u$. By Lemma 1.2.1, all multihomogeneous components of any element from I are contained in I. Thus $u \in I \implies \operatorname{lin} u \in I$. The converse follows from the formula

$$u(z_1, \ldots, z_n) = \frac{1}{m_1! \ldots m_n!}(\operatorname{lin} u)(\underbrace{z_1, \ldots, z_1}_{m_1}, \ \cdots, \ \underbrace{z_n, \ldots, z_n}_{m_n}). \quad \square$$

Notes. 1. If the ground ring K is arbitrary then, as in the previous statement, the proof of Lemma 1.2.2 remains valid when $1/(m-1)! \in K$.

2. Lemma 1.2.2 contains several results from [24], namely Theorem 1, Corollaries 3, 4 and 5.

A variety of group representations is said to be a *Magnus variety* if it is generated by stable representations. If $\mathcal{X}$ is a Magnus variety then, by 0.2.6, all free representations of $\mathcal{X}$ belong to the class VSC$\mathcal{K}$, where $\mathcal{K}$ is the class of all stable representations from $\mathcal{X}$. It follows that every free representation $\phi = (E, F)$ of $\mathcal{X}$ is *residually stable*, that is, E possesses a system of F-submodules A_i such that $\bigcap A_i = 0$ and the corresponding factor-representations $\phi_i = (E/A_i, F)$ are stable. Conversely, if all free representations of $\mathcal{X}$ are residually stable, then $\mathcal{X}$ is certainly generated by stable representations and therefore is Magnus.

1.2.3. Proposition. *A variety $\mathcal{X}$ is homogeneous and Magnus if and only if for arbitrary $u \in KF$*

$$u \in \operatorname{Id} \mathcal{X} \iff \forall n : u_{(n)} \in \operatorname{Id} \mathcal{X}. \tag{2}$$

P r o o f. Let $\mathcal{X}$ be a homogeneous Magnus variety. Then $\implies$ is automatically valid and it suffices to prove the reverse implication. Denote $\operatorname{Id} \mathcal{X} = I$ and let u be an element of KF such that $u_{(n)} \in I$ for all n. Since $\mathcal{X}$ is a Magnus variety, the representation $\operatorname{Fr} \mathcal{X} = (KF/I, F)$ is residually stable. In other words, the following condition is satisfied in the algebra $\overline{KF} = KF/I$:

$$\bigcap_{k=1}^{\infty} \bar{\Delta}^k = \bar{0}$$

or, equivalently,

$$\bigcap_{k=1}^{\infty} (I + \Delta^k) = I \tag{3}$$

where Δ is the augmentation ideal of KF. Present u in the form $u = \sum_{i=1}^{k-1} u_{(i)} + \sum_{i=k}^{\infty} u_{(i)}$. The first summand belongs to I by the choice of u. The second summand belongs to Δ^k by Proposition 0.5.4 (an element of KF is contained in Δ^k if and only if its homogeneous components of degrees $< n$ are all equal to zero). Hence $u \in I + \Delta^k$ for each k and so, by (3), $u \in I$, as required.

Conversely, let $\mathcal{X}$ satisfy (2). Since $\Longrightarrow$ means homogeneity, it is enough to show that $\mathcal{X}$ is a Magnus variety, i.e. that (3) is satisfied for $I = \operatorname{Id} \mathcal{X}$. Let $u \in \bigcap_{k=1}^{\infty}(I + \Delta^k)$, then for each k we have $u = v_k + w_k$, where $v_k \in I$, $w_k \in \Delta^k$. Hence for $i = 1, \ldots, k-1$ the i-th homogeneous component of u coincides with the i-th homogeneous component of v_k. Since $v_k \in I$ and I is homogeneous, all the homogeneous components of v_k belong to I. Thus $u_{(1)}, \ldots, u_{(k-1)} \in I$, and this is true for arbitrary k. It remains to apply $\Longleftarrow$. $\square$

Let M be a set of identities. It is natural to ask what properties of the identities from M will guarantee that the variety determined by M is homogeneous? For example, is it enough if the elements of M are themselves homogeneous? The answer to this question will be given in § 1.5, but now we establish the following fact.

1.2.4. Proposition (Grinberg and Krop [24]). *Any variety determined by a set of multilinear identities is homogeneous.*

Proof. Let M be a set of multilinear identities and $I = \operatorname{Id}(M)$ the verbal ideal of KF generated by M. We have to show that I is homogeneous.

a) Let us prove that if $u(z_1, \ldots, z_n) \in M$ and $c_1, \ldots, c_n$ are monomials in arbitrary variables z_i, then $u(c_1, \ldots, c_n) \in I$. Without loss of generality we may assume that the monomials $c_1, \ldots, c_n$ involve mutually disjoint sets of variables z_i and, furthermore, that all these variables are different from $z_1, \ldots, z_n$.

First we prove that $u(c_1, z_2, \ldots, z_n) \in I$. We proceed by induction on $\deg c_1$. If $\deg c_1 = 1$, it is nothing to prove. Suppose that our assertion has been proved for $\deg c_1 < m$, and let $\deg c_1 = m$, that is, $c_1 = z_{i_1} z_{i_2} \ldots z_{i_m}$. Let φ be the endomorphism of the group F defined by the rule:

$$x_1^\varphi = x_{i_1} x_{i_2} \ldots x_{i_m}, \qquad x_j^\varphi = x_j \quad \text{for} \quad j \neq 1.$$

Then

$$
\begin{aligned}
z_1^\varphi = (x_1 - 1)^\varphi &= x_{i_1} x_{i_2} \ldots x_{i_m} - 1 \\
&= (1 + z_{i_1})(1 + z_{i_2}) \ldots (1 + z_{i_m}) - 1 \\
&= c_1 + \sum d_k
\end{aligned}
$$

where each d_k is a monomial of degree $< m$ in variables z_i. Since the word $u(z_1, \ldots, z_n)$ is linear with respect to z_1, we have

$$
\begin{aligned}
I \ni u(z_1, \ldots, z_n)^\varphi &= u(z_1^\varphi, \ldots, z_n^\varphi) \\
&= u(c_1 + \sum d_k, \ z_2, \ldots, z_n) \\
&= u(c_1, z_2, \ldots, z_n) + \sum u(d_k, z_2, \ldots, z_n).
\end{aligned}
\tag{4}
$$

Each summand $u(d_k, z_2, \ldots, z_n)$ belongs to I by the induction hypothesis, therefore it follows from (4) that $u(c_1, z_2, \ldots, z_n) \in I$, as desired.

Repeating this argument, we successively obtain that $u(c_1, c_2, z_3, \ldots \ldots, z_n) \in I$, $u(c_1, c_2, c_3, \ldots, z_n) \in I$, etc. (It is essential that u is multilinear and that each successive monomial involves "new" variables z_{i_k}.)

b) Let now $w \in I$; we have to prove that $w_{(m)} \in I$ for every m. Since w is a finite sum of elements of the form $a u^\varphi b$, where $a, b \in KF$, $u \in M$ and $\varphi \in \operatorname{End} F$, we may assume without loss of generality that $w = a u^\varphi b$. Let

$$u = u(z_1, \ldots, z_n), \qquad z_j^\varphi = c^j = \sum_{i=1}^\infty c_{(i)}^j \quad (j = 1, \ldots, n),$$

$$a = \sum_{i=0}^\infty a_{(i)}, \qquad b = \sum_{i=0}^\infty b_{(i)}$$

(note that the homogeneous component $c^j_{(0)}$ of c^j must be equal to 0 because $z_j \in \Delta$ and therefore $c^j = z^\varphi_j \in \Delta$ as well). Then

$$w = \sum a_{(i)} \cdot u(\sum c^1_{(i)}, \ldots, \sum c^n_{(i)}) \cdot \sum b_{(i)}.$$

Using multilinearity of u, we see that the homogeneous component $w_{(m)}$ of degree m of w is a finite sum of elements of the form

$$a_{(k)} u(c^1_{(i_1)}, \ldots, c^n_{(i_n)}) b_{(l)} \tag{5}$$

where $k + i_1 + \cdots + i_n + l = m$. We emphasize that $a_{(k)}, c^j_{(i_j)}, b_{(l)}$ are homogeneous polynomials of degrees k, i_j, l respectively. Further, since the word $u(z_1, \ldots, u_n)$ is multilinear and each $c^j_{(i_j)}$ is a sum of monomials, we eventually obtain that $w_{(m)}$ is a finite sum of elements of the form

$$a_{(k)} u(d_1, \ldots, d_n) b_{(l)}$$

where all d_i are monomials. By the first part of the proof, $u(d_1, \ldots, d_n) \in I$ whence the theorem follows. $\square$

The results of this section lead naturally to the following questions. First, surprisingly, we do not have any examples of homogeneous non-Magnus varieties.

1.2.5. Problem. *Is every homogeneous variety Magnus?*

To explain the essence of this problem, recall that a variety $\mathcal{X}$ is homogeneous if and only if for any $u \in KF$

$$u \in \operatorname{Id} \mathcal{X} \implies \forall n : u_{(n)} \in \operatorname{Id} \mathcal{X}. \tag{6}$$

It remains to compare (6) and (2).

1.2.6. Problem. *Is every variety definable by multilinear identities Magnus?*

In view of Theorem 1.2.4, the connection between the above two problems is obvious. Note that if Problem 1.2.5 is solved in the affirmative, then

Theorem 1.2.4 can be reversed, at least over a field of characteristic zero. Indeed, let $\operatorname{char} K = 0$, I a homogeneous fully invariant ideal of KF and M the set of all multilinear elements from I. If u is an arbitrary element from I, then $u_{(n)} \in I$ for all n and so $\operatorname{lin} u_{(n)} \in I$ by Lemma 1.2.2. Hence $\operatorname{lin} u_{(n)} \in M \subseteq \operatorname{Id}(M)$. The ideal $\operatorname{Id}(M)$ is homogeneous by 1.2.4, therefore, applying 1.2.2 again, we see that $u_{(n)} \in \operatorname{Id}(M)$. Now suppose that the answer to (i) is positive. Then $\operatorname{Id}(M)$ is a homogeneous Magnus verbal ideal and, by 1.2.3, we have $u \in \operatorname{Id}(M)$. Thus $I = \operatorname{Id}(M)$, that is, I is generated by its multilinear words.

1.3. Connections with varieties of associative algebras

One connection between varieties of (associative) algebras over K and varieties of group representations over K is obvious. Let $\mathcal{M}$ be a variety of algebras over K and let $T = T(\mathcal{M})$ be the set of all its identities in the free associative algebra $K\langle X \rangle$ (without 1!) on a countable set of variables $X = \{x_1, x_2, \ldots\}$. This set is a T-ideal of $K\langle X \rangle$, that is, an ideal which is admissible under all endomorphism of the K-algebra $K\langle X \rangle$. Now, since $K\langle X \rangle$ is naturally contained in the group algebra KF, T is a subset of KF and therefore determines a variety of group representations over K, say $\mathcal{M}'$.

However, the connection $\mathcal{M} \mapsto \mathcal{M}'$ does not have really deep consequences. The point is that there is an essential difference between identities of associative algebras (=polynomial identities) and identities of group representations. Consider, for example, two simple polynomial identities

$$x_1 x_2 \ldots x_n \quad \text{and} \quad x^n.$$

They determine the important varieties of n-nilpotent algebras and n-nil-algebras respectively. But as identities of group representations they determine *nothing* (or more formally, each of them determines the variety $\mathcal{E}$ of zero representations), because we can substitute 1 for all the x_i!

The aim of the present section is to establish another connection between varieties of algebras and varieties of group representations, which proves to be more interesting and efficient. We will systematically use the canonical embedding of algebras

$$K\langle Z\rangle \subset KF \subset K\langle\langle Z\rangle\rangle$$

under which $x_i = 1 + z_i$ are free generators of the group F. For brevity and convenience of formulations, the variety $\mathcal{E}$ of zero representations is excluded from the subsequent considerations.

1.3.1. Theorem (Vovsi [98]). *Let K be an infinite field. Then to every variety $\mathcal{M}$ of associative algebras over K there canonically corresponds a homogeneous Magnus variety $\mathcal{M}^\alpha$ of group representations over K. Moreover:*

(a) *the map α is injective;*

(b) *if char $K = 0$, then $\operatorname{Im}\alpha$ coincides with the set of all homogeneous Magnus varieties of group representations over K;*

(c) *a variety of algebras $\mathcal{M}$ is nilpotent of class n if and only if $\mathcal{M}^\alpha$ is stable of class n;*

(d) *α is a monomorphism of semigroups, that is, $(\mathcal{M}_1\mathcal{M}_2)^\alpha = \mathcal{M}_1^\alpha\mathcal{M}_2^\alpha$ for all varieties of algebras $\mathcal{M}_1$ and $\mathcal{M}_2$;[1]*

(e) *α is a homomorphism of lattices, that is, $(\mathcal{M}_1 \vee \mathcal{M}_2)^\alpha = \mathcal{M}_1^\alpha \vee \mathcal{M}_2^\alpha$ and $(\mathcal{M}_1 \wedge \mathcal{M}_2)^\alpha = \mathcal{M}_1^\alpha \wedge \mathcal{M}_2^\alpha$ for all varieties of algebras $\mathcal{M}_1$ and $\mathcal{M}_2$.*

P r o o f. 1) For an arbitrary T-ideal T of $K\langle Z\rangle$, let

$$T^\alpha = \{u \in KF | \forall n : u_{(n)} \in T\}. \tag{1}$$

In other words, T^α can be defined as follows. Consider in $K\langle Z\rangle$ the filtration

$$K\langle Z\rangle \supset (K\langle Z\rangle)^2 \supset \cdots \supset (K\langle Z\rangle)^n \supset \cdots,$$

then the completion $\widehat{K\langle Z\rangle}$ of $K\langle Z\rangle$, corresponding to this filtration, is precisely the algebra $K\langle\langle Z\rangle\rangle$. If now $\hat{T}$ is the completion of T, then $T^\alpha = \hat{T} \cap KF$.

---

[1]Here the product of varieties of algebras is meant in the sense of Bergman–Lewin [6], that is, $\mathcal{M}_1\mathcal{M}_2$ is the variety corresponding to the T-ideal $T_2 T_1$ where $T_i = T(\mathcal{M}_i)$.

We prove that T^α is a homogeneous Magnus verbal ideal of KF. First, it is clear that T^α is a proper K-subspace of KF. Second, if $u = \sum u_{(n)} \in T^\alpha$, $v = \sum v_{(n)} \in KF$, then $(uv)_{(n)} = \sum_{k+l=n} u_{(k)} v_{(l)}$. Since $u_{(k)} \in T$ for all k, it follows that $(uv)_{(n)} \in T$ for all n. Therefore $uv \in T^\alpha$ and similarly $vu \in T^\alpha$. Thus T^α is an ideal of KF. Let us show that T^α is invariant under an arbitrary endomorphism φ of F. Again let $u = u(z_1, \ldots, u_m) = \sum u_{(n)} \in T^\alpha$, then $u_{(n)} \in T$ for each n. We have to prove that if $v = u^\varphi$, then $v_{(n)} \in T$ for each n. Fix an arbitrary n and, calculating modulo Δ^{n+1}, obtain

$$v = u^\varphi \equiv \Big(\sum_{i=1}^{n} u_{(i)} \Big)^\varphi \equiv \sum_{i=1}^{n} u_{(i)}(z_1^\varphi, \ldots, z_m^\varphi) \mod \Delta^{n+1}.$$

Modulo Δ^{n+1}, each z_j^φ is a finite polynomial in z_j, say $z_j^\varphi = f_j(z_1, \ldots, z_k)$, $j = 1, \ldots, n$. Thus

$$v \equiv v_{(1)} + \cdots + v_{(n)} \equiv \sum_{i=1}^{n} u_{(i)}(f_1, \ldots, f_m) \mod \Delta^{n+1}.$$

Now we note that since $u_{(i)}(z_1, \ldots, z_m) \in T$, it follows that $u_{(i)}(f_1, \ldots, f_m) \in T$ as well, whence $v_{(1)} + \cdots + v_{(n)} \in T$. This is true for all n, and hence $v_{(n)} \in T$ for each n, as required.

Thus T^α is a verbal ideal of KF. It follows from (1) and Proposition 1.2.3 that it is homogeneous and Magnus. Finally, since all T-ideals over an infinite field are homogeneous, it is clear from (1) that the map $T \mapsto T^\alpha$ is injective. In terms of varieties we can say that α is an injection of the set of varieties of algebras into the set of homogeneous Magnus varieties of group representations.

2) Now we will prove that if char $K = 0$ and I is a homogeneous Magnus verbal ideal of KF, then the set

$$I^\beta = \{ u \in K\langle Z \rangle \mid \forall n : u_{(n)} \in I \} \tag{2}$$

is a T-ideal of $K\langle Z \rangle$. Clearly I^β is an ideal, and since I is multihomogeneous (Lemma 1.2.1), it is evident that I^β is multihomogeneous as well. Therefore it suffices to prove that if $u = u(z_1, \ldots, z_k)$ is a multihomogeneous element from I^β and $\varphi \in \mathrm{End}(K\langle Z \rangle)$, then $u^\varphi \in I^\beta$. Denote

$$\deg_{z_i} u = m_i, \quad m_1 + \cdots + m_k = m.$$

Then $u = u_{(m)}$ and, by (2), $u \in I$. Let $v = v(z_1, \ldots, z_m) = \operatorname{lin} u$. By Lemma 1.2.2, $v \in I$ and, again by (2), we obtain $v \in I^\beta$. Let us show that $v^\varphi \in I^\beta$.

(i) Let first all z_i^φ be monomials:

$$z_1^\varphi = z_{11} \ldots z_{1r_1}, \, z_2^\varphi = z_{21} \ldots z_{2r_2}, \, \ldots, \, z_m^\varphi = z_{m1} \ldots z_{mr_m}.$$

Then $v^\varphi = v(z_{11} \ldots z_{1r_1}, \ldots, z_{m1} \ldots z_{mr_m})$. Our objective is to obtain the same element from v by means of *group* endomorphisms. Denote $r_1 + \cdots + r_m = r$ and apply to v an endomorphism π of the group F such that

$$x_1^\pi = x_{11} \ldots x_{1r_1}, x_2^\pi = x_{21} \ldots x_{2r_2}, \ldots, x_m^\pi = x_{m1} \ldots x_{mr_m}.$$

It is easy to verify that

$$z_i^\pi = z_{i1} \ldots z_{ir_i} + \quad \text{lower degree terms}.$$

Therefore, it follows from the multilinearity of v that

$$I \ni v^\pi = v(z_1^\pi, \ldots, z_m^\pi)$$
$$= v(z_{11} \ldots z_{1r_1}, \ldots, z_{m1} \ldots z_{mr_m}) + \quad \text{lower degree terms}.$$

The last summand is exactly v^φ and simultaneously it is the homogeneous component of degree r of v^π. Since I is homogeneous, we have $v^\varphi \in I$, whence $v^\varphi \in I^\beta$.

(ii) Now let $z_i^\varphi = f_i$ be arbitrary polynomials from $K\langle Z \rangle$. Then $f_i = \sum_j \alpha_{ij} f_{ij}$, where $\alpha_{ij} \in K$ and f_{ij} are monomials. Since v is multilinear, it follows that

$$v^\varphi = v(f_1, \ldots, f_m) = \sum {}_{j_1, \ldots, j_m} \alpha_{1j_1} \ldots \alpha_{mj_m} v(f_{1j_1}, \ldots, f_{mj_m}).$$

All the $v(f_{1j_1}, \ldots, f_{mj_m})$ belong to I^β by (i), therefore $v^\varphi \in I^\beta$.

Now we return to the word $u = u(z_1, \ldots, z_k)$. It is a consequence, in the sense of ring identities, of its full linearization v, that is,

$$u = \sum a_i v^{\psi_i} b_i \quad \text{where} \quad a_i, b_i \in K\langle Z \rangle, \, \psi_i \in \operatorname{End}(K\langle Z \rangle).$$

Then $u^\varphi = \sum a_i^\varphi v^{\psi_i \varphi} b_i^\varphi$ and by the above we have $v^{\psi_i \varphi} \in I^\beta$ for all i. Therefore $u^\varphi \in I^\beta$ and so I^β is a T-ideal of $K\langle Z \rangle$.

3) Let $\operatorname{char} K = 0$. It follows from 1), 2) and Proposition 1.2.3 that if T is a T-ideal of $K\langle Z \rangle$, then $T^{\alpha\beta} = T$, but if I is a homogeneous Magnus verbal ideal of KF, then $I^{\beta\alpha} = I$. Thus the maps α and β realize a one-to-one correspondence between all T-ideals of $K\langle Z \rangle$ and all homogeneous Magnus verbal ideals of KF. In other words, over a field of characteristic zero, α is a bijection of the set of all varieties of algebras onto the set of all homogeneous Magnus varieties of group representations.

4) To prove that a variety of algebras $\mathcal{M}$ is n-nilpotent if and only if the variety of representations $\mathcal{M}^\alpha$ is n-stable, it suffices to show that

$$z_1 z_2 \ldots z_n \in T \iff z_1 z_2 \ldots z_n \in T^\alpha$$

for any T-ideal T of $K\langle Z \rangle$. But this is evident in view of (1) and (2).

5) It remains to prove that α is a monomorphism of semigroups and lattices. Let T_1 and T_2 be arbitrary T-ideals of the algebra $K\langle Z \rangle$. We prove that $(T_1 T_2)^\alpha = T_1^\alpha T_2^\alpha$. Let $w \in (T_1 T_2)^\alpha$, that is, $w \in KF$ and $w_{(n)} \in T_1 T_2$ for each n. This means that for each n

$$w_{(n)} = \sum_{i=1}^{k_i} u_i^n v_i^n, \quad \text{where} \quad u_i^n \in T_1, \ v_i^n \in T_2$$

(do not confuse the upper indices with exponents!). Since T-ideals over an infinite field are homogeneous, it follows from (1) that $T_i \subseteq T_i^\alpha$, whence $u_i^n \in T_1^\alpha$, $v_i^n \in T_2^\alpha$ ($i = 1, \ldots, k_n$) and so $w_{(n)} \in T_1^\alpha T_2^\alpha$. It is known that the product of Magnus varieties of group representations is also a Magnus variety [80, Theorem 24.3.2], hence $T_1^\alpha T_2^\alpha$ is a Magnus verbal ideal. Proposition 1.2.3 implies now that $w \in T_1^\alpha T_2^\alpha$.

Conversely, let $w \in T_1^\alpha T_2^\alpha$, that is,

$$w = u^1 v^1 + \cdots + u^m v^m, \quad \text{where} \quad u^i \in T_1^\alpha, \ v_i \in T_2^\alpha.$$

By the definition of the map α, we have $u_{(n)}^i \in T_1$, $v_{(n)}^i \in T_2$ for all i and n. Therefore

$$w_{(n)} = \sum_{i=1}^{m} (u^i v^i)_{(n)} = \sum_{i=1}^{m} \left(\sum_{k+l=n} u_{(k)}^i v_{(l)}^i \right) \in T_1 T_2$$

and hence $w \in (T_1 T_2)^\alpha$. Thus $T_1^\alpha T_2^\alpha = (T_1 T_2)^\alpha$.

The proof of the equality $(T_1 \cap T_2)^\alpha = T_1^\alpha \cap T_2^\alpha$ is even simpler. Indeed,

$$w \in (T_1 \cap T_2)^\alpha \iff \forall n: w_{(n)} \in T_1 \cap T_2 \iff$$

$$\iff (\forall n: w_{(n)} \in T_1) \ \& \ (\forall n: w_{(m)} \in T_2) \iff (w \in T_1^\alpha) \ \& \ (w \in T_2^\alpha).$$

In a similar way, one can verify that $(T_1 + T_2)^\alpha = T_1^\alpha + T_2^\alpha$. This completes the proof. $\square$

Note. In proving Theorem 1.3.1, we deliberately restricted ourselves to representations over infinite fields where the result has a quite complete form. However, it should be noted that some parts of the theorem are valid over any ground field and even over an arbitrary commutative ring.

Theorem 1.3.1 has several applications. First, it implies the following interesting fact.

1.3.2. Corollary. *Let* char $K = 0$. *Then for every n the lattice of all n-nilpotent varieties of algebras over K is isomorphic to the lattice of homogeneous n-stable varieties of group representations over K.* $\square$

Let us have a look at this isomorphism from the standpoint of identities. If $\mathcal{M}$ is a variety of algebras and $\mathcal{M}^\alpha$ the corresponding variety of group representations then, by (1) and (2), the identities of these varieties have the same homogeneous components. The only principal difference consists in the fact that identities of $\mathcal{M}$ are (finite!) polynomials in z_i, but the identities of $\mathcal{M}^\alpha$ are (infinite!) series in z_i. If now $\mathcal{M}$ is a n-nilpotent variety of algebras, then $\mathcal{M}^\alpha$ is n-stable and therefore $\mathrm{Id}(\mathcal{M}^\alpha) \supseteq \Delta^n$. This shows that, modulo Δ^n, elements from $\mathrm{Id}(\mathcal{M}^\alpha)$ can also be regarded as finite polynomials. In other words, *modulo the word $z_1 z_2 \ldots z_n$, the identities of $\mathcal{M}$ and the identities of $\mathcal{M}^\alpha$ are just the same.*

Thus n-stable varieties are uniquely and *explicitly* determined by T-ideals of the free associative algebra $K\langle Z \rangle = A$ which contain the ideal A^n. Since all such ideals are homogeneous, but every homogeneous word is equivalent to a multilinear word of the same degree, these ideals are characterized by their multilinear elements. More exactly, let Π_k be the subspace of A consisting of all multilinear polynomials

$$u(z_1,\ldots,z_k) = \sum_{\sigma \in S_k} \lambda_\sigma z_{\sigma(1)} \ldots z_{\sigma(k)}$$

of degree k from A. If T is a T-ideal of A containing A^n, then denote $T_k = T \cap \Pi_k$ $(k = 1,\ldots,n-1)$. Evidently:

(i) T_k is a subspace of Π_k invariant under the natural action of the symmetric group S_k;

(ii) if $u = u(z_1,\ldots,z_k) \in T_k$, then $uz_{k+1} \in T_{k+1}, z_{k+1}u \in T_{k+1}, u(z_1,\ldots, z_{i-1}, z_i z_{k+1}, z_{i+1},\ldots, z_k) \in T_{k+1}$ and $u(z_1,\ldots, z_{i-1}, z_{k+1}z_i, z_{i+1},\ldots \ldots, z_k) \in T_{k+1}$.

It is commonly known that the correspondence $T \mapsto (T_1, T_2, \ldots, T_{n-1})$ is a bijection between all T-ideals of the algebra A containing A^n, and all $(n-1)$-tuples $(T_1, T_2, \ldots, T_{n-1})$ satisfying (i)–(ii). This leads to the following result.

1.3.3. Corollary (Grinberg and Krop [24]). *Let* char $K = 0$. *Then there exists a one-to-one correspondence between all $(n-1)$-tuples $(T_1, T_2, \ldots, T_{n-1})$ satisfying (i)-(ii) and all n-stable homogeneous varieties of group representations over K.* □

In particular, take a variety $\mathcal{X}$ of group representations such that $S^{n-1} \subseteq \mathcal{X} \subseteq S^n$. Then $\Delta^{n-1} \supseteq \operatorname{Id} \mathcal{X} \supseteq \Delta^n$ (whence $\mathcal{X}$ is certainly homogeneous), and so in the corresponding $(n-1)$-tuple $(T_1, T_2, \ldots, T_{n-1})$ we have $T_1 = \cdots = T_{n-2} = 0$. Therefore $\operatorname{Id} \mathcal{X}$ is completely determined by its subspace T_{n-1} (plus, of course, the word $z_1 z_2 \ldots z_n$), and we obtain another result from [24]:

1.3.4. Corollary. *Let* char $K = 0$. *Then there exists a one-to-one correspondence between varieties $\mathcal{X}$ of group representations such that $S^{n-1} \subseteq \mathcal{X} \subseteq S^n$, and S_{n-1}-invariant subspaces of Π_{n-1}.* □

We note in conclusion that homogeneous varieties of group representations were first considered in [23] and [24]. Most of the results obtained there were analogous to well known facts of the theory of varieties of associative algebras, but the roots of this analogy remained unclear. Theorem 1.3.1 clarifies the situation completely. In addition, it shows that the collection $\mathbb{M}(K)$ of varieties of group representations is really very rich. From § 0.2 we already know that to each variety of groups one can assign a variety of group representations $\omega\Theta$, and that the map $\Theta \mapsto \omega\Theta$ is injective. It is natural to say that varieties of representations $\omega\Theta$ are of *group type*. Now we have constructed the injective map $\mathcal{M} \mapsto \mathcal{M}^{\alpha}$ (provided K is an infinite field) from the set of varieties of associative algebras over K to $\mathbb{M}(K)$; it is natural to say that varieties of $\mathcal{M}^{\alpha}$ are of *ring type*. Thus the problems of studying varieties of groups and associative algebras are "embeddable" in the corresponding problem for varieties of group representations.

1.3.5. Problem. *Describe all varieties of representations which are simultaneously of group type and of ring type.*

If, in particular, char $K = 0$ then, by Theorem 1.3.1 (b), this is equivalent to describing all varieties of groups Θ such that $\omega\Theta$ is homogeneous and Magnus simultaneously.

1.4. Stable varieties of class 4

In the present section, following [23], we will give a classification of 4-stable varieties of group representations over a field of characteristic zero. We start with one auxiliary statement which in an explicit form was established in [23], but in essence had been proved much earlier (see for instance [54] and [39]). Recall that $\mathcal{U}_2$ is the variety of 2-unipotent representations, i.e. the variety determined by the identity $(x - 1)^2$.

1.4.1. Proposition. *Over an arbitrary ring with $1/2$, the variety $\mathcal{U}_2$ is stable of class 3.*

P r o o f. Denote $U_2 = \operatorname{Id} \mathcal{U}_2$ and let $\equiv$ denote equality modulo U_2 in the algebra KF. We have to prove that $z_1 z_2 z_3 \equiv 0$ where, as usual, $z_i = x_i - 1$.

Apply the endomorphism $x_1 \mapsto x_1 x_2$ of F to the equality

$$z_1^2 \equiv 0. \tag{1}$$

Then $z_1 \mapsto z_1 + z_2 + z_1 z_2$, and it follows from (1) that

$$0 \equiv (z_1 + z_2 + z_1 z_2)^2 \equiv z_1 z_2 + z_2 z_1 + z_1 z_2 z_1 + z_2 z_1 z_2 + z_1 z_2 z_1 z_2.$$

Multiplying the last equality by z_1 on the left and by z_2 on the right, we obtain $z_1 z_2 z_1 z_2 \equiv 0$, whence

$$z_1 z_2 + z_2 z_1 + z_1 z_2 z_1 + z_2 z_1 z_2 \equiv 0.$$

Again multiply by z_1 on the left; we get $z_1 z_2 z_1 \equiv 0$ and so $z_2 z_1 z_2 \equiv 0$. Consequently,

$$z_1 z_2 \equiv -z_2 z_1. \tag{2}$$

Now apply to (1) the endomorphism $x_1 \mapsto x_1 x_2 x_3$. Then $z_1 \mapsto z_1 + z_2 + z_3 + z_1 z_2 + z_1 z_3 + z_2 z_3 + z_1 z_2 z_3$, and it follows from (1) and (2) that

$$\begin{aligned}
0 &\equiv (z_1 + z_2 + z_3 + z_1 z_2 + z_1 z_3 + z_2 z_3 + z_1 z_2 z_3)^2 \\
&\equiv 2 z_1 z_2 z_3 + z_1 z_3 z_2 + z_3 z_1 z_2 + z_2 z_3 z_1 + z_2 z_1 z_3 \\
&\equiv 2 z_1 z_2 z_3 + (z_1 z_3 + z_3 z_1) z_2 + z_2 (z_3 z_1 + z_1 z_3) \\
&\equiv 2 z_1 z_2 z_3.
\end{aligned}$$

Since $1/2 \in K$, it follows that $z_1 z_2 z_3 \equiv 0$. $\square$

1.4.2. Theorem (Grinberg [23]). *Over an arbitrary field of characteristic $\neq 2$, every 4-stable variety is homogeneous.*

Proof. Let $\mathcal{X}$ be a 4-stable variety, $I = \operatorname{Id} \mathcal{X}$ and $w \in I$. It has to be proved that $w_{(n)} \in I$ for each n. Since $I \supseteq \Delta^4$, we may assume that $w = w_{(1)} + w_{(2)} + w_{(3)}$.

Recall that a word $v \in KF$ is said to be *normal* if all monomials of its decomposition into a formal power series involve the same set of variables z_i; this set is called the *support* of v and is denoted by $\operatorname{Supp} v$. Each word from KF can be uniquely written as a sum of a finite number of normal words which are consequences of the initial word.

We return to the word $w \in I$. Present w as a sum of normal words. Since these words are consequences of w, they all belong to I. Therefore, without any loss of generality, we may assume that $w = w_{(1)} + w_{(2)} + w_{(3)}$ is a normal word from I. Since the monomials of w have degrees ≤ 3, it follows that $|\mathrm{Supp}\, w| \leq 3$.

If $|\mathrm{Supp}\, w| = 3$, then $w = w_{(3)}$ and there is nothing to prove. If $|\mathrm{Supp}\, w| = 1$, then $w = \alpha z_1 + \beta z_1^2 + \gamma z_1^3$, where $\alpha, \beta, \gamma \in K$. As usual, denote by $\equiv$ equality modulo I. Then

$$\alpha z_1 + \beta z_1^2 + \gamma z_1^3 \equiv 0.$$

If $\alpha \neq 0$, we multiply the last formula by z_1^2 and, since $z_1^4 \equiv 0$, it follows that $\alpha z_1^3 \equiv 0$. Hence $z_1^3 \equiv 0$ and $\alpha z_1 + \beta z_1^2 \equiv 0$. Multiplying the latter by z_1, we obtain $z_1^2 \equiv 0$, whence $z_1 \equiv 0$. Thus $w_{(1)} \equiv w_{(2)} \equiv w_{(3)} \equiv 0$. If $\alpha = 0$, that is $w = \beta z_1^2 + \gamma z_1^3$, the argument is similar.

It remains to consider the case when $|\mathrm{Supp}\, w| = 2$. Then $w_{(1)} = 0$ and

$$w = w_{(2)} + w_{(3)} \equiv 0. \tag{3}$$

Here $w_{(2)} = \alpha z_1 z_2 + \beta z_2 z_1$ and $w_{(3)} = w' + w''$, where w' is homogeneous of degree 1 in z_1 and of degree 2 in z_2, and w'' is homogeneous of degree 2 in z_1 and of degree 1 in z_2.

a) Let $\alpha + \beta \neq 0$. Applying the map $z_2 \mapsto z_1$ to (3), we obtain

$$(\alpha + \beta)z_1^2 + \gamma z_1^3 \equiv 0$$

whence, by the same argument as above, $z_1^2 \equiv 0$. Therefore it follows from Proposition 1.4.1 that $\Delta^3 \subseteq I$, and so $w_{(3)} \equiv 0$ and $w_{(2)} = w - w_{(3)} \equiv 0$.

b) Let $\alpha + \beta = 0$. Then $w_{(2)} = \alpha(z_1 z_2 - z_2 z_1)$ and, since for $\alpha = 0$ there is nothing to prove, we may assume that $\alpha = 1$. Thus

$$w = z_1 z_2 - z_2 z_1 + w' + w'' \equiv 0. \tag{4}$$

Let φ be the endomorphism of F defined as follows:

$$x_1^\varphi = x_1^{-1},$$
$$x_i^\varphi = x_i \quad \text{if} \quad i \neq 1.$$

Then $z_1^\varphi = -z_1 + z_1^2 - z_1^3 + \ldots$, $z_2^\varphi = z_2$ and, since $\Delta^4 \subseteq I$, it follows that $w'^\varphi \equiv -w'$, $w''^\varphi \equiv w''$. Applying φ to (4), we obtain

$$0 \equiv (-z_1 + z_1^2 - \ldots)z_2 - z_2(-z_1 + z_1^2 - \ldots) - w' + w''$$
$$\equiv -z_1 z_2 + z_2 z_1 + z_1^2 z_2 - z_2 z_1^2 - w' + w'',$$

and since $z_1^2 z_2 - z_2 z_1^2 \equiv z_1 w + w z_1 \equiv 0$, it follows that

$$-z_1 z_2 + z_2 z_1 - w' + w'' \equiv 0. \tag{5}$$

Adding (4) and (5), we see that $2w'' \equiv 0$, whence $w'' \equiv 0$ and similarly $w' \equiv 0$. Thus $w_{(3)} = w' + w'' \equiv 0$ and therefore $w_{(2)} \equiv 0$. $\square$

Consider several applications. Until the end of this section we take K to be a field of characteristic zero. Denote by $L(\mathcal{S}^4)$ the lattice of all subvarieties of $\mathcal{S}^4$, by $\mathcal{N}_4$ the variety of nilpotent algebras of class 4 over K, and by $L(\mathcal{N}_4)$ the lattice of all subvarieties of $\mathcal{N}_4$. Then Corollary 1.3.2 and Theorem 1.4.2 immediately imply:

1.4.3. Corollary. $L(\mathcal{S}^4) \simeq L(\mathcal{N}_4)$. $\square$

This implies, in particular, that the lattice $L(\mathcal{S}^4)$ is not distributive, since it is well known that $L(\mathcal{N}_4)$ is not distributive — see [2].

It is now easy to describe all 4-stable varieties. According to 1.3.3 and 1.4.2, these varieties are in one-to-one correspondence with 3-tuples (T_1, T_2, T_3) satisfying the conditions (i)–(ii) from the previous section. Namely, to each 3-tuple (T_1, T_2, T_3) there corresponds the variety detrmined by the set of identities $T_1 \cup T_2 \cup T_3 \cup \{z_1 z_2 z_3 z_4\}$. It remains to describe all such 3-tuples.

1. If $T_1 \neq \{0\}$, then $T_1 = \Pi_1 = \{\alpha z_1 \mid \alpha \in K\}$. It follows from (ii) that $T_2 = \Pi_2$ and $T_3 = \Pi_3$.

2. Let $T_1 = \{0\}$. Then T_2 can be an arbitrary $K S_2$-submodule of the (right) $K S_2$-module $\Pi_2 = \{\alpha z_1 z_2 + \beta z_2 z_1 \mid \alpha, \beta \in K\}$. Every proper $K S_2$-submodule of Π_2 is one-dimensional, and it is easy to see that there are only two such submodules:

$$A = \{\alpha(z_1 z_2 - z_2 z_1) \mid \alpha \in K\}, \quad B = \{\beta(z_1 z_2 + z_2 z_1) \mid \beta \in K\}.$$

Therefore T_2 must be equal to one of the following: $\{0\}, A, B, \Pi_2$.

a) $T_2 = \Pi_2$. Then $T_3 = \Pi_3$.

b) $T_2 = \{0\}$. Then T_3 can be an arbitrary KS_3-submodule of Π_3.

c) $T_2 = B$. Then $z_1 z_2 + z_2 z_1$ is an identity of the corresponding variety, and so z_1^2 is an identity of this variety as well. By 1.4.1, $z_1 z_2 z_3 \in T_3$, whence $T_3 = \Pi_3$.

d) $T_2 = A$. Let C be a subset of Π_3 consisting of all elements

$$\sum_{\sigma \in S_3} \alpha_\sigma z_{\sigma(1)} z_{\sigma(2)} z_{\sigma(3)}$$

such that $\sum \alpha_\sigma = 0$. Evidently C is a KS_3-submodule of Π_3, and since $\dim_K \Pi_3 = 6$ but $\dim_K C = 5$, we see that C is a maximal KS_3submodule of Π_3. Furthermore, since $(\{0\}, A, T_3)$ satisfies (ii), it is easy to understand that $T_3 \supseteq C$. Consequently, either $T_3 = C$ or $T_3 = \Pi_3$.

Thus the description of all tuples (T_1, T_2, T_3) satisfying (i)–(ii) is finished. This completes the description of all 4-stable varieties of group representations. For the sake of clarity, we list the above 3-tuples together with bases of identities of the corresponding varieties and their standard notation (if such a notation exists):

(T_1, T_2, T_3)	*Basis of identities*	*Notation*
(Π_1, Π_2, Π_3)	z_1	$\mathcal{S}$
$(\{0\}, \Pi_2, \Pi_3)$	$z_1 z_2$	$\mathcal{S}^2$
$(\{0\}, B, \Pi_3)$	$z_1 z_2 + z_2 z_1$ (or z_1^2)	$\mathcal{U}_2$
$(\{0\}, A, C)$	$z_1 z_2 - z_2 z_1,\ z_1 z_2 z_3 z_4$	$\omega \mathcal{A} \wedge \mathcal{S}^4$
$(\{0\}, A, \Pi_3)$	$z_1 z_2 - z_2 z_1,\ z_1 z_2 z_3$	$\omega \mathcal{A} \wedge \mathcal{S}^3$
$(\{0\}, \{0\}, T_3)$	$T_3 \cup \{z_1 z_2 z_3 z_4\}$	

Here T_3 runs over the set of all KS_3-submodules of Π_3; it is known that the cardinality of this set is equal to the cardinality of the ground field K. Thus, there exist five "sporadic" 4-stable varieties and one series of cardinality $|K|$ consisting of varieties intermediate between $\mathcal{S}^3$ and $\mathcal{S}^4$.

A careful reader will notice that in the above list the trivial variety $\mathcal{E}$ is missing. This is not accidental, for at the very beginning of §1.3 we excluded this variety from considerations.

1.5. A stable nonhomogeneous variety

Assume, for the present, that char $K = 0$. In this situation Theorem 1.3.1 applies with full force and yields a substantially complete picture. But new questions continue to appear.

1. Is every stable variety homogeneous?

2. Is it true that every variety, definable by homogeneous identities, is homogeneous?

3. Is it true that if a variety of algebras $\mathcal{M}$ is defined by identities $\{u_i(z_1, \ldots, z_n)\}$, then the variety of representations $\mathcal{M}^\alpha$ is defined by the identities $\{u_i(x_1 - 1, \ldots, x_n - 1)\}$?

Partial answers to these questions are given by Proposition 1.2.4 and Theorem 1.4.2, but in their general form they remained open for about ten years.[2] Especially intriguing, in view of the results of §1.3, was the first question. Indeed, if it were answered in the affirmative then, by Corollary 1.3.2, the study of stable varieties of group representations in zero characteristic would be completely reduced to the study of nilpotent varieties of algebras. Fortunately for the theory of varieties of group representations, each of the above questions has a negative answer.

1.5.1. Theorem (Vovsi [98]). *Over an arbitrary field of characteristic $\neq 2$, the variety $\mathcal{X}_0$ defined by the identities*

$$u(x_1, x_2) = 2z_1 z_2 z_1 - z_1 z_2^2 z_1, \quad v(x_1, x_2, x_3, x_4, x_5) = z_1 z_2 z_3 z_4 z_5$$

does not satisfy the identity $2z_1 z_2 z_1$ and is definable by homogeneous identities.

Since $\mathcal{X}_0$ satisfies v, it is stable of class 5. Since $\mathcal{X}_0$ satisfies u but does not satisfy the homogeneous component of degree 3 of u, it is not

---

[2] Moreover, it was not even known whether a variety, which is *simultaneously* stable and definable by homogeneous identities, is necessarily homogeneous (see [24, p. 22]).

homogeneous. This answers questions 1 and 2. Now consider the variety of algebras $\mathcal{M}$ defined by u and v. Then the variety of representations $\mathcal{M}^\alpha$ must be homogeneous, whence $\mathcal{M}^\alpha \neq \mathcal{X}_0$. This yields a negative answer to the last question.

The proof of Theorem 1.5.1 consists of a number of lemmas. We begin by proving its last assertion.

1.5.2. Lemma. *The variety $\mathcal{X}_0$ can be defined by the homogeneous identities*

$$z_1^3, \quad z_1 z_2^2 + z_2^2 z_1 - z_1^2 z_2 - z_2 z_1^2, \quad z_1 z_2^2 + z_2^2 z_1 + 2 z_1 z_2 z_1,$$
$$z_1 z_2 z_1 z_3, \quad z_3 z_1 z_2 z_1, \quad z_1^2 z_2^2 - z_2^2 z_1^2, \quad z_1^2 z_2^2 + z_1 z_2^2 z_1, \quad z_1 z_2 z_3 z_4 z_5. \tag{1}$$

P r o o f. We show first that the identities (1) are satisfied in $\mathcal{X}_0$. The symbol $\equiv$ will denote equality modulo $\operatorname{Id} \mathcal{X}_0$. Substituting x_1 for x_2 in $u(x_1, x_2)$, we have $2z_1^3 - z_1^4 \equiv 0$. Multiplying this equation by z_1, we get $z_1^4 \equiv 0$ and, consequently, $z_1^3 \equiv 0$. Multiplying $u(x_1, x_2)$ by z_3 on the right, we get $z_1 z_2 z_1 z_3 \equiv 0$, and by z_3 on the left, $z_3 z_1 z_2 z_1 \equiv 0$.

Now we apply the endomorphism $x_1 \mapsto x_1 x_2$ to $u(x_1, x_2)$ and use the equations just derived. We obtain

$$0 \equiv 2(z_1 + z_2 + z_1 z_2) z_2 (z_1 + z_2 + z_1 z_2) - (z_1 + z_2 + z_1 z_2) z_2^2 (z_1 + z_2 + z_1 z_2)$$

$$\equiv 2(z_1 z_2 z_1 + z_2^2 z_1 + z_1 z_2^2 z_1 + z_1 z_2^2) - z_1 z_2^2 z_1$$

$$\equiv (2 z_1 z_2 z_1 - z_1 z_2^2 z_1) + 2(z_2^2 z_1 + z_1 z_2^2 + z_1 z_2^2 z_1).$$

Consequently,

$$z_1 z_2^2 + z_2^2 z_1 + z_1 z_2^2 z_1 \equiv 0. \tag{2}$$

¿From (2) and the identity $2 z_1 z_2 z_1 - z_1 z_2^2 z_1 \equiv 0$, we have $z_1 z_2^2 + z_2^2 z_1 + 2 z_1 z_2 z_1 \equiv 0$. Multiplying by z_1 on the left, we have $z_1^2 z_2^2 + z_1 z_2^2 z_1 \equiv 0$, and by z_1 on the right, $z_2^2 z_1^2 + z_1 z_2^2 z_1 \equiv 0$. In the latter we transpose z_1 and z_2 and compare the equation obtained with the next-to-last equation, which yields $z_1 z_2^2 z_1 \equiv z_2 z_1^2 z_2$. But then, according to (2), we have $z_1 z_2^2 + z_2^2 z_1 - z_1^2 z_2 - z_2 z_1^2 \equiv 0$. Furthermore, it is now clear that $z_1^2 z_2^2 - z_2^2 z_1^2 \equiv 0$. Thus, the identities (1) are satisfied in $\mathcal{X}_0$.

Conversely, let us show that the identities (1) imply the identity $u(x_1, x_2)$. Let I be the verbal ideal of KF generated by the words (1), and let now $\equiv$ denote the equality modulo I. Applying the endomorphism $x_1 \mapsto x_1 x_2$ to $z_1^3 \equiv 0$ and using formulas (1), we get

$$
\begin{aligned}
0 &\equiv (z_1 + z_2 + z_1 z_2)^3 \equiv (z_1 + z_2 + z_1 z_2)^2 (z_1 + z_2 + z_1 z_2) \\
&\equiv (z_1^2 + z_2 z_1 + z_1 z_2 z_1 + z_1 z_2 + z_2^2 + z_1 z_2^2 + z_1^2 z_2 + z_2 z_1 z_2)(z_1 + z_2 + z_1 z_2) \\
&\equiv z_2 z_1^2 + z_1 z_2 z_1 + z_2^2 z_1 + z_1 z_2^2 z_1 + z_1^2 z_2 + z_2 z_1 z_2 + z_1 z_2^2 + z_1^2 z_2^2 + z_2 z_1^2 z_2 \\
&\equiv (z_1^2 z_2 + z_2 z_1^2 + z_2 z_1^2 z_2) + (z_2^2 z_1 + z_1 z_2^2 + z_1 z_2^2 z_1) + \\
&\qquad\qquad\qquad\qquad\qquad\qquad + (z_1 z_2 z_1 + z_2 z_1 z_2 + z_1^2 z_2^2) \\
&\equiv (-2 z_2 z_1 z_2 + z_2 z_1^2 z_2) + (-2 z_1 z_2 z_1 + z_1 z_2^2 z_1) + (z_1 z_2 z_1 + z_2 z_1 z_2 + z_1^2 z_2^2) \\
&\equiv (*).
\end{aligned}
$$

It follows from (1) that $z_1 z_2 z_1 \equiv z_2 z_1 z_2$ and $-z_1^2 z_2^2 \equiv z_1 z_2^2 z_1 \equiv z_2 z_1^2 z_2$. Applying these equations together with (1) to $(*)$, we see that

$$
(*) \equiv -u(x_1, x_2) - u(x_1, x_2) + u(x_1, x_2) = -u(x_1, x_2)
$$

and so $u(x_1, x_2) \equiv 0$. $\square$

Let V be the space of dimension 10 over K with basis $e_0, e_1, \ldots, e_9$. Take in $M_{10}(K)$ two matrices

$$
a_1 = 1 + e_{12} + e_{24} + e_{36} - \frac{1}{2} e_{59} - \frac{1}{2} e_{50} + e_{60} - e_{78} + e_{79} + e_{70} - e_{89} - e_{80},
$$

$$
a_2 = 1 + e_{13} + e_{25} + e_{37} + e_{49} + e_{58} - \frac{1}{2} e_{69} - \frac{1}{2} e_{60} + e_{99} + e_{90} - e_{09} - e_{00},
$$

where 1 is the identity matrix and the e_{ij} are the usual matrix units. It is easy to see that the matrices a_1 and a_2 are invertible. Let $G = \langle a_1, a_2 \rangle$ be the subgroup they generate in $GL_{10}(K)$ and let $\rho = (V, G)$ be the natural representations of G. Theorem 1.5.1 will be completely proved if we establish that ρ satisfies the identities u and v, but does not satisfy $2 z_1 z_2 z_1$.

1.5.3. Lemma. *The representation ρ satisfies the identity v, that is, ρ is 5-stable.*

Proof. For an arbitrary subset M in V, let $\langle M \rangle$ be its linear span. Consider the following series of subspaces of V:

$$\langle 0 \rangle \subset \langle e_9 + e_0 \rangle \subset \langle e_8, e_9, e_{10} \rangle \subset \langle e_4, e_5, e_6, e_7, e_8, e_9, e_{10} \rangle \subset V.$$

A straightforward verification shows that all terms of this series are invariant with respect to a_1 and a_2, and on all its factors a_1 and a_2 act as the identity. Therefore ρ is 5-stable. $\square$

All subsequent calculations are carried out in the algebra $M_{10}(K)$. For an arbitrary matrix $g \in M_{10}(K)$, set $\bar{g} = g - 1$. It follows from Lemma 1.5.3 that

$$\bar{g}_1 \bar{g}_2 \bar{g}_3 \bar{g}_4 \bar{g}_5 = 0 \qquad (3)$$

for any $g_i \in G$. We also note that if $w = w(x_1, \ldots, x_n)$ is an element of KF, then the substitution of matrices $g_i \in G$ for variables x_i in w is equivalent to the substitution of the matrices $\bar{g}_i$ for the differences $z_i = x_i - 1$.

1.5.4. Lemma. *The following equations are true:*

$$\bar{a}_1^2 = e_{14} + e_{30} + e_{79} + e_{70}, \qquad (4)$$

$$\bar{a}_2^2 = e_{17} + e_{28} + e_{49} + e_{40}, \qquad (5)$$

$$\bar{a}_1 \bar{a}_2 = e_{15} + e_{29} - \frac{1}{2} e_{39} - \frac{1}{2} e_{30} - e_{69} - e_{60}, \qquad (6)$$

$$\bar{a}_2 \bar{a}_1 = e_{16} - \frac{1}{2} e_{29} - \frac{1}{2} e_{20} - e_{38} + e_{39} + e_{30} - e_{59} - e_{50}, \qquad (7)$$

$$\bar{a}_1^3 = \bar{a}_2^3 = 0, \qquad (8)$$

$$\bar{a}_1 \bar{a}_2 \bar{a}_1 = \bar{a}_2 \bar{a}_1 \bar{a}_2 = -\frac{1}{2} e_{19} - \frac{1}{2} e_{10}, \qquad (9)$$

$$\bar{a}_1 \bar{a}_2^2 = e_{18} + e_{29} + e_{20}, \quad \bar{a}_2^2 \bar{a}_1 = -e_{18} + e_{19} + e_{10} - e_{29} - e_{20}, \qquad (10)$$

$$\bar{a}_2 \bar{a}_1^2 = e_{10} + e_{39} + e_{30}, \quad \bar{a}_1^2 \bar{a}_2 = -e_{19} - e_{39} - e_{30}, \qquad (11)$$

$$\bar{a}_1 \bar{a}_2^2 \bar{a}_1 = \bar{a}_2 \bar{a}_1^2 \bar{a}_2 = -e_{19} - e_{10}, \qquad (12)$$

$$\bar{a}_i \bar{a}_j \bar{a}_i \bar{a}_k = \bar{a}_j \bar{a}_i \bar{a}_k \bar{a}_i \quad \text{for any} \quad i, j, k \in \{1, 2\}, \qquad (13)$$

$$\bar{a}_1^2 \bar{a}_2^2 = \bar{a}_2^2 \bar{a}_1^2 = e_{19} + e_{10}. \qquad (14)$$

Proof can be done by direct calculation. For example,

$$\bar{a}_1\bar{a}_2 = \left(e_{12} + e_{24} + e_{36} - \frac{1}{2}e_{59} - \frac{1}{2}e_{50} + e_{60} - e_{78} + e_{79} + e_{70} - e_{89} - e_{80}\right).$$

$$\cdot\left(e_{13} + e_{25} + e_{37} + e_{49} + e_{58} - \frac{1}{2}e_{69} - \frac{1}{2}e_{60} + e_{99} + e_{90} - e_{09} - e_{00}\right) =$$

$$= e_{15} + e_{29} - \frac{1}{2}e_{39} - \frac{1}{2}e_{30} - \frac{1}{2}e_{59} + e_{79} - e_{89} - \frac{1}{2}e_{50} + e_{70} - e_{80} +$$

$$+\frac{1}{2}e_{59} - e_{69} - e_{70} + e_{89} + \frac{1}{2}e_{50} - e_{60} - e_{70} + e_{80} =$$

$$= e_{15} + e_{29} - \frac{1}{2}e_{39} - \frac{1}{2}e_{30} - e_{69} - e_{60}. \quad \square$$

1.5.5. Corollary. *The representation ρ does not satisfy the identity* $2z_1z_2z_1$.

Indeed, according to (9) we have $2\bar{a}_1\bar{a}_2\bar{a}_1 = -e_{19} - e_{10} \neq 0$. $\quad \square$

1.5.6. Lemma. *Let*

$$w(x_1, x_2, x_3, x_4) = \sum_{\sigma \in S_4} \alpha_\sigma z_{\sigma(1)} z_{\sigma(2)} z_{\sigma(3)} z_{\sigma(4)}$$

be a word from KF multilinear in z_1, z_2, z_3, z_4. This word is an identity of ρ if and only if it vanishes on the generators a_1 and a_2 of G, i.e. when $w(a_i, a_j, a_k, a_l) = 0$ for any $i, j, k, l \in \{1, 2\}$.

Proof. Assume that the condition of the lemma is satisfied, and let $g_1, g_2, g_3, g_4 \in G$. It suffices to show that $w(g_1, g_2, g_3, g_4) = 0$. Each g_i is a product of the elements $a_1, a_2, a_1^{-1}, a_2^{-1}$:

$$g_i = a_{i1}^{\epsilon_{i1}} a_{i2}^{\epsilon_{i2}} \dots a_{ik_i}^{\epsilon_{ik_i}} \quad (i = 1, 2, 3, 4; \; \epsilon_{ij} = \pm 1, \; a_{ij} \in \{a_1, a_2\}).$$

Using the formulas

$$\overline{gh} = \bar{g} + \bar{h} + \bar{g}\bar{h}, \quad \overline{a_i^{-1}} = -\bar{a}_i + \bar{a}_i^2 \; (i = 1, 2) \tag{15}$$

(the latter follows from (8)), it is not hard to see that $\bar{g}_i = \epsilon_{i1}\bar{a}_{i1} + \cdots + \epsilon_{ik_i}\bar{a}_{ik_i} +$ monomials of degree ≥ 2 in $\bar{a}_1, \bar{a}_2$. Therefore from the multilinearity of w it follows that

$$w(g_1, g_2, g_3, g_4) = \sum_{\sigma \in S_4} \alpha_\sigma \bar{g}_{\sigma(1)} \bar{g}_{\sigma(2)} \bar{g}_{\sigma(3)} \bar{g}_{\sigma(4)}$$

$$= \sum \epsilon_{1p} \epsilon_{2q} \epsilon_{3r} \epsilon_{4s} w(a_{1p}, a_{2q}, a_{3r}, a_{4s}) +$$

$$+ \text{ monomials of degree} \geq 5 \text{ in } \bar{a}_1, \bar{a}_2.$$

Here $w(a_{1p}, a_{2q}, a_{3r}, a_{4s}) = 0$ by the condition, but monomials of degree ≥ 5 are equal to 0 according to (3). Therefore $w(g_1, g_2, g_3, g_4) = 0$. □

1.5.7. Lemma. *The following multilinear words are identities of the representation ρ :*

$$w_1(x_1, x_2, x_3, x_4) = z_1 z_2 z_3 z_4 + z_3 z_2 z_1 z_4,$$

$$w_2(x_1, x_2, x_3, x_4) = z_1 z_2 z_3 z_4 + z_1 z_4 z_3 z_2,$$

$$w_3(x_1, x_2, x_3, x_4) = z_1 z_2 z_3 z_4 + z_3 z_4 z_2 z_1 + z_4 z_2 z_1 z_3 + z_3 z_2 z_1 z_4,$$

$$w_4(x_1, x_2, x_3, x_4) = z_1 z_2 z_3 z_4 - z_1 z_3 z_2 z_4 + z_4 z_2 z_3 z_1 - z_4 z_3 z_2 z_1,$$

$$w_5(x_1, x_2, x_3, x_4) = z_1 z_4 z_2 z_3 + z_4 z_1 z_2 z_3 + z_3 z_2 z_1 z_4 - z_3 z_2 z_4 z_1 +$$

$$+ 2 z_1 z_3 z_2 z_4 + 2 z_4 z_3 z_2 z_1.$$

Proof. Since the proofs of all five cases are similar, we restrict ourselves to w_1. In view of Lemma 1.5.6, it is sufficient to prove that $w_1(a_i, a_j, a_k, a_l) = 0$ for any $i, j, k, l \in \{1, 2\}$. If some three of the indices i, j, k, l are pairwise equal then, by (13), all monomials of the expression $w_1(a_i, a_j, a_k, a_l)$ vanish. Therefore we may assume that two of the indices i, j, k, l equal 1, and the remaining two equal 2. Then six cases are possible:

1) $i = j = 1$, $k = l = 2$,   4) $i = j = 2$, $k = l = 1$,

2) $i = k = 1$, $j = l = 2$,   5) $i = k = 2$, $j = l = 1$,

3) $i = l = 1$, $j = k = 2$,   6) $i = l = 2$, $j = k = 1$.

Using the identities (12)–(14), in the first three cases we have

$$w_1(a_i, a_j, a_k, a_l) = \bar{a}_1^2 \bar{a}_2^2 + \bar{a}_2 \bar{a}_1^2 \bar{a}_2 = 0,$$

$$w_1(a_i, a_j, a_k, a_l) = \bar{a}_1 \bar{a}_2 \bar{a}_1 \bar{a}_2 + \bar{a}_1 \bar{a}_2 \bar{a}_1 \bar{a}_2 = 0,$$

$$w_1(a_i, a_j, a_k, a_l) = \bar{a}_1 \bar{a}_2^2 \bar{a}_1 + \bar{a}_2^2 \bar{a}_1^2 = 0.$$

The remaining three cases are obtained from those already examined by transposing the indices 1 and 2. It follows from (12)–(14) that under this transposition the values of the corresponding monomials do not change: $\bar{a}_i \bar{a}_j \bar{a}_k \bar{a}_l = \bar{a}_{\tau(i)} \bar{a}_{\tau(j)} \bar{a}_{\tau(k)} \bar{a}_{\tau(l)}$ where $\tau = $ (12). Thus, also in these three cases $w_1(a_i, a_j, a_k, a_l) = 0$. □

1.5.8. Corollary. *The words* $z_1 z_2 z_1 z_3$, $z_1 z_2 z_3 z_2$, $z_1^2 z_2^2 + z_2 z_1^2 z_2$, $z_1^2 z_2^2 + z_1 z_2^2 z_1$ *are identities of the representation* ρ.

Proof. It suffices to apply the previous lemma and the obvious equalities $z_1 z_2 z_1 z_3 = \frac{1}{2} w_1(x_1, x_2, x_1, x_3)$, $z_1 z_2 z_3 z_2 = \frac{1}{2} w_2(x_1, x_2, x_3, x_2)$, $z_1^2 z_2^2 + z_1 z_2^2 z_1 = w_2(x_1, x_1, x_2, x_2)$, $z_1^2 z_2^2 + z_2 z_1^2 z_2 = w_1(x_1, x_1, x_2, x_2)$. □

1.5.9. Lemma. *Let* $w_6(x_1, x_2, x_3) = 2(z_1 z_2 z_3 + z_3 z_2 z_1 + z_1 z_3 z_2 z_1 + z_3 z_2 z_1 z_3) - z_1 z_2^2 z_3 - z_3 z_2^2 z_1$. *Then* $w_6(a, b, c) = 0$ *for arbitrary* $a, b, c \in \{a_1, a_2, a_1^{-1}, a_2^{-1}\}$.

Proof. 1. We show that $w_6(a, b, c) = 0$ for any $a, b, c \in \{a_1, a_2\}$. If all three elements a, b, c are equal to the same a_i, then $w_6(a, b, c) = 0$ by (8). It remains to consider the cases when two of the elements a, b, c coincide with some a_i and the third equals $a_j \neq a_i$. There are three such cases: $a = b \neq c$, $a = c \neq b$, $a \neq b = c$. Using Lemma 1.5.4, we analyze these cases respectively:

a) $w_6(a_i, a_i, a_j) = 2(\bar{a}_i^2 \bar{a}_j + \bar{a}_j \bar{a}_i^2 + \bar{a}_i \bar{a}_j \bar{a}_i^2 + \bar{a}_j \bar{a}_i^2 \bar{a}_j) -$
$- \bar{a}_j^3 \bar{a}_i - \bar{a}_i \bar{a}_j^3 = 2(\bar{a}_i^2 \bar{a}_j + \bar{a}_j \bar{a}_i^2 + \bar{a}_j \bar{a}_i^2 \bar{a}_j) =$
$= 2(e_{19} + e_{10} - e_{19} - e_{10}) = 0,$

b) $w_6(a_i, a_j, a_i) = 2(\bar{a}_i \bar{a}_j \bar{a}_i + \bar{a}_i \bar{a}_j \bar{a}_i + \bar{a}_i^2 \bar{a}_j \bar{a}_i + \bar{a}_i \bar{a}_j \bar{a}_i^2) -$
$- \bar{a}_i \bar{a}_j^2 \bar{a}_i - \bar{a}_i \bar{a}_j^2 \bar{a}_i = 4\bar{a}_i \bar{a}_j \bar{a}_i - 2\bar{a}_i \bar{a}_j^2 \bar{a}_i =$
$= -2e_{19} - 2e_{10} + 2e_{19} + 2e_{10} = 0,$

c) $w_6(a_j, a_i, a_i) = 2(\bar{a}_j \bar{a}_i^2 + \bar{a}_i^2 \bar{a}_j + \bar{a}_j \bar{a}_i^2 \bar{a}_j + \bar{a}_i^2 \bar{a}_j \bar{a}_i) - \bar{a}_j \bar{a}_i^3 - \bar{a}_i^3 \bar{a}_j =$
$= 2(\bar{a}_j \bar{a}_i^2 + \bar{a}_i^2 \bar{a}_j + \bar{a}_j \bar{a}_i^2 \bar{a}_j) =$
$= 2(e_{19} + e_{10} - e_{19} - e_{10}) = 0.$

2. Let us show that for arbitrary $a, b, c \in \{a_1, a_2\}$

$$w_6(a^{-1}, b, c) = w_6(a, b^{-1}, c) = w_6(a, b, c^{-1}) = 0. \qquad (16)$$

By Lemmas 1.5.4, 1.5.7 and formulas (3) and (15), we have:

a) $\ w_6(a^{-1}, b, c) = 2[(-\bar{a} + \bar{a}^2)\bar{b}\bar{c} + \bar{c}\bar{b}(-\bar{a} + \bar{a}^2) + \bar{a}\bar{c}\bar{b}\bar{a}-$

$\qquad - \bar{c}\bar{b}\bar{a}\bar{c}] + \bar{a}\bar{b}^2\bar{c} + \bar{c}\bar{b}^2\bar{a} = [-2(\bar{a}\bar{b}\bar{c} + \bar{c}\bar{b}\bar{a} + \bar{a}\bar{c}\bar{b}\bar{a} + \bar{c}\bar{b}\bar{a}\bar{c}) + \bar{a}\bar{b}^2\bar{c}+$

$\qquad \bar{c}\bar{b}^2\bar{a}] + 2(\bar{a}^2\bar{b}\bar{c} + \bar{c}\bar{b}\bar{a}^2 + 2\bar{a}\bar{c}\bar{b}\bar{a}) =$

$\qquad = -w_6(a, b, c) + w_5(a, b, c, a) = 0,$

b) $\ w_6(a, b^{-1}, c) = 2[\bar{a}(-\bar{b} + \bar{b}^2)\bar{c} + \bar{c}(-\bar{b} + \bar{b}^2)\bar{a} + \bar{a}\bar{c}(-\bar{b} + \bar{b}^2)\bar{a}+$

$\qquad + \bar{c}(-\bar{b} + \bar{b}^2)\bar{a}\bar{c}] - \bar{a}(-\bar{b} + \bar{b}^2)^2 - \bar{c}(-\bar{b} + \bar{b}^2)^2\bar{a} =$

$\qquad = -2(\bar{a}\bar{b}\bar{c} + \bar{c}\bar{b}\bar{a} + \bar{a}\bar{c}\bar{b}\bar{a} + \bar{c}\bar{b}\bar{a}\bar{c}) + \bar{a}\bar{b}^2\bar{c} + \bar{c}\bar{b}^2\bar{a} =$

$\qquad = -w_6(a, b, c) = 0,$

c) $\ w_6(a, b, c^{-1}) = 2[\bar{a}\bar{b}(-\bar{c} + \bar{c}^2) + (-\bar{c} + \bar{c}^2)\bar{b}\bar{a} + \bar{a}(-\bar{c} + \bar{c}^2)\bar{b}\bar{a}+$

$\qquad + (-\bar{c} + \bar{c}^2)] - \bar{a}\bar{b}^2(-\bar{c} + \bar{c}^2) - (\bar{c} + \bar{c}^2)\bar{b}^2\bar{a} = [-2(\bar{a}\bar{b}\bar{c} + \bar{c}\bar{b}\bar{a}+$

$\qquad + \bar{a}\bar{c}\bar{b}\bar{a} + \bar{c}\bar{b}\bar{a}\bar{c}) + \bar{a}\bar{b}^2\bar{c} + \bar{c}\bar{b}^2\bar{a}] + 2(\bar{a}\bar{b}\bar{c}^2 + \bar{c}^2\bar{b}\bar{a} + 2\bar{c}\bar{b}\bar{a}\bar{c}) =$

$\qquad = -w_6(a, b, c) + w_5(c, b, a, c) + 2w_4(c, b, a, c) = 0.$

Thus (16) is proved. The proof of the lemma can now be finished by a simple combination of the examined cases 1 and 2. $\square$

1.5.10. Lemma. $w_6(a, g, h) = 0$ for any $a \in \{a_1, a_2, a_1^{-1}, a_2^{-1}\}$ and $g, h \in G$.

Proof. 1. We show that $w_6(a, b, h) = 0$ for any $a, b \in \{a_1, a_2, a_1^{-1}, a_2^{-1}\}$, $h \in G$. Denote by $l(h)$ the length of a shortest presentation of h as a product of the elements $a_i^{\pm 1}$ $(i = 1, 2)$, and apply induction on $l(h)$. If $l(h) = 1$, then everything follows from Lemma 1.5.9. Let $l(h) = n$, and suppose that for elements of shorter length the required fact is proved. Present h in the form $h = h_1 h_2$, where $l(h_i) < n$. Then

$$w_6(a, b, h) = 2[\bar{a}\bar{b}(\bar{h}_1 + \bar{h}_2 + \bar{h}_1\bar{h}_2) + (\bar{h}_1 + \bar{h}_2 + \bar{h}_1\bar{h}_2)\bar{b}\bar{a}+$$

$$+ \bar{a}(\bar{h}_1 + \bar{h}_2 + \bar{h}_1\bar{h}_2)\bar{b}\bar{a} + (\bar{h}_1 + \bar{h}_2 + \bar{h}_1\bar{h}_2)\bar{b}\bar{a}(\bar{h}_1+$$

$$+ \bar{h}_2 + \bar{h}_1\bar{h}_2)] - \bar{a}\bar{b}^2(\bar{h}_1 + \bar{h}_2 + \bar{h}_1\bar{h}_2) - (\bar{h}_1 + \bar{h}_2\bar{h}_1\bar{h}_2)\bar{b}^2\bar{a}$$

$$= 2(\bar{a}\bar{b}\bar{h}_1 + \bar{a}\bar{b}\bar{h}_2 + \bar{a}\bar{b}\bar{h}_1\bar{h}_2 + \bar{h}_1\bar{b}\bar{a} + \bar{h}_2\bar{b}\bar{a} + \bar{h}_1\bar{h}_2\bar{b}\bar{a} +$$
$$+ \bar{a}\bar{h}_1\bar{b}\bar{a} + \bar{a}\bar{h}_2\bar{b}\bar{a} + \bar{h}_1\bar{b}\bar{a}\bar{h}_2 + \bar{h}_2\bar{b}\bar{a}\bar{h}_1 +$$
$$+ \bar{h}_2\bar{b}\bar{a}\bar{h}_2) - \bar{a}\bar{b}^2\bar{h}_1 - \bar{a}\bar{b}^2\bar{h}_2 - \bar{h}_1\bar{b}^2\bar{a} - \bar{h}_2\bar{b}^2\bar{a}$$
$$= w_6(a, b, h_1) + w_6(a, b, h_2) + \bar{a}\bar{b}\bar{h}_1\bar{h}_2 + \bar{h}_1\bar{h}_2\bar{b}\bar{a} + \bar{h}_1\bar{b}\bar{a}\bar{h}_2 + \bar{h}_2\bar{b}\bar{a}\bar{h}_1$$
$$= w_6(a, b, h_1) + w_6(a, b, h_2) + w_3(a, b, h_1, h_2).$$

Since $w_6(a, b, h_1) = w_6(a, b, h_2) = 0$ by the induction hypothesis and $w_3(a, b, h_1, h_2) = 0$ by Lemma 1.5.7, it follows that $w_6(a, b, h) = 0$.

2. Now we can prove the lemma in general, applying induction on $l(g)$. If $l(g) = 1$, the proof follows from the above. Assume that $l(g) = n$ and for elements of shorter length the lemma is proved. Then, presenting g in the form $g = g_1 g_2$, where $l(g_i) < n$, we have

$$w_6(a, g, h) = 2[\bar{a}(\bar{g}_1 + \bar{g}_2 + \bar{g}_1\bar{g}_2)\bar{h} + \bar{h}(\bar{g}_1 + \bar{g}_2 + \bar{g}_1\bar{g}_2)\bar{a} +$$
$$+ \bar{a}\bar{h}(\bar{g}_1 + \bar{g}_2 + \bar{g}_1\bar{g}_2)\bar{a} + \bar{h}(\bar{g}_1 + \bar{g}_2 + \bar{g}_1\bar{g}_2)\bar{a}\bar{h}] -$$
$$- \bar{a}(\bar{g}_1 + \bar{g}_2 + \bar{g}_1\bar{g}_2)^2\bar{h} - \bar{h}(\bar{g}_1 + \bar{g}_2 + \bar{g}_1\bar{g}_2)^2\bar{a}$$
$$= w_6(a, g_1, h) + w_6(a, g_2, h) + 2\bar{a}\bar{g}_1\bar{g}_2\bar{h} + 2\bar{h}\bar{g}_1\bar{g}_2\bar{a} -$$
$$- \bar{a}\bar{g}_1\bar{g}_2\bar{h} - \bar{a}\bar{g}_2\bar{g}_1\bar{h} - \bar{h}\bar{g}_1\bar{g}_2\bar{a} - \bar{h}\bar{g}_2\bar{g}_1\bar{a}$$
$$= w_6(a, g_1, h) + w_6(a, g_2, h) + w_4(a, g_1, g_2, h) = 0. \quad \square$$

1.5.11. Lemma. *The word* $u(x_1, x_2) = 2z_1 z_2 z_1 - z_1 z_2^2 z_1$ *is an identity of the representation* ρ.

Proof. 1. We show that $u(a, b) = 0$ for any $a, b \in \{a_1, a_2, a_1^{-1}, a_2^{-1}\}$. It follows directly from (4)–(14) that $u(a_1, a_1) = u(a_1, a_2) = u(a_2, a_1) = u(a_2, a_2) = 0$. Further, for any $i, j \in \{1, 2\}$

$$u(a_i, a_j) = 2\bar{a}_i(-\bar{a}_j + \bar{a}_j^2)\bar{a}_i - \bar{a}_i(-\bar{a}_j + \bar{a}_j^2)^2\bar{a}_i$$
$$= -2\bar{a}_i\bar{a}_j\bar{a}_i + 2\bar{a}_i\bar{a}_j^2\bar{a}_i - \bar{a}_i\bar{a}_j^2\bar{a}_i$$
$$= -u(a_i, a_j) = 0,$$
$$u(a_i^{-1}, a_j) = 2(-\bar{a}_i + \bar{a}_i^2)\bar{a}_j(-\bar{a}_i + \bar{a}_i^2) - (-\bar{a}_i + \bar{a}_i^2)\bar{a}_j^2\bar{a}_i + \bar{a}_i^2)$$
$$= 2(\bar{a}_i\bar{a}_j\bar{a}_i - \bar{a}_i\bar{a}_j\bar{a}_i^2 - \bar{a}_i^2\bar{a}_j\bar{a}_i) - \bar{a}_i\bar{a}_j^2\bar{a}_i$$
$$= 2\bar{a}_i\bar{a}_j\bar{a}_i - \bar{a}_i\bar{a}_j^2\bar{a}_i = u(a_i, a_j) = 0.$$

Similarly, $u(a_i^{-1}, a_j^{-1}) = -u(a_i, a_j) = 0$.

2. We show that $u(a, h) = 0$ for any $a \in \{a_1, a_2, a_1^{-1}, a_2^{-1}\}$, $h \in G$. For $l(h) = 1$ we use the previous argument. If $h = h_1 h_2$, where $l(h_i) < l(h)$, then

$$u(a, h) = 2\bar{a}(\bar{h}_1 + \bar{h}_2 + \bar{h}_1\bar{h}_2)\bar{a} - \bar{a}(\bar{h}_1 + \bar{h}_2 + \bar{h}_1\bar{h}_2)^2\bar{a}$$
$$= 2(\bar{a}\bar{h}_1\bar{a} + \bar{a}\bar{h}_2\bar{a} + \bar{a}\bar{h}_1\bar{h}_2\bar{a}) - \bar{a}\bar{h}_1^2\bar{a} - \bar{a}\bar{h}_1\bar{h}_2\bar{a} - \bar{a}\bar{h}_2\bar{h}_1\bar{a} - \bar{a}\bar{h}_2^2\bar{a}$$
$$= u(a, h_1) + u(a, h_2) + \frac{1}{2}w_4(a, h_1, h_2, a).$$

The proof is completed by induction.

3. Now let us show by induction on $l(g)$ that $u(g, h) = 0$ for any $g, h \in G$. If $l(g) = 1$, this follows from the above. Assume that $l(g) = n$ and that for elements of shorter length the lemma is proved. Then g can be written in the form $g = af$, where $a \in \{a_1, a_2, a_1^{-1}, a_2^{-1}\}$ and $l(f) = n - 1$. Consequently,

$$u(g, h) = 2(\bar{a} + \bar{f} + \bar{a}\bar{f})\bar{h}(\bar{a} + \bar{f} + \bar{a}\bar{f}) - (\bar{a} + \bar{f} + \bar{a}\bar{f})\bar{h}^2(\bar{a} + \bar{f} + \bar{a}\bar{f})$$
$$= 2(\bar{a}\bar{h}\bar{a} + \bar{f}\bar{h}\bar{a} + \bar{a}\bar{f}\bar{h}\bar{a} + \bar{a}\bar{h}\bar{f} + \bar{f}\bar{h}\bar{f} + \bar{a}\bar{f}\bar{h}\bar{f} + \bar{a}\bar{h}\bar{a}\bar{f} +$$
$$+ \bar{f}\bar{h}\bar{a}\bar{f}) - (\bar{a}\bar{h}^2\bar{a} + \bar{a}\bar{h}^2\bar{f} + \bar{f}\bar{h}^2\bar{a} + \bar{f}\bar{h}^2\bar{f})$$
$$= u(a, h) + u(f, h) + [2(\bar{a}\bar{h}\bar{f} + \bar{f}\bar{h}\bar{a} + \bar{a}\bar{f}\bar{h}\bar{a} + \bar{f}\bar{h}\bar{a}\bar{f}) -$$
$$- \bar{a}\bar{h}^2\bar{f} - \bar{f}\bar{h}^2\bar{a}] + 2\bar{a}\bar{f}\bar{h}\bar{f} + 2\bar{a}\bar{h}\bar{a}\bar{f}$$
$$= u(a, h) + u(f, h) + w_6(a, h, f) + 2\bar{a}\bar{f}\bar{h}\bar{f} + 2\bar{a}\bar{h}\bar{a}\bar{f}.$$

Applying the induction hypothesis, Lemma 1.5.10 and Corollary 1.5.8, we eventually obtain $u(g, h) = 0$. $\square$

Theorem 1.5.1 follows directly from 1.5.2, 1.5.3, 1.5.5 and 1.5.11. $\square$

It is interesting to note that the variety $\mathcal{X}_0$ is, in a sense, a *minimal* example of nonhomogeneous stable variety. To make this assertion more precise, suppose that I is a nonhomogeneous verbal ideal which is stable of class n (that is, $I \supseteq \Delta^n$). Then there exists $u = \sum_{i=0}^{\infty} u_{(i)} \in I$ such that $u_{(i)} \notin I$ for some i. Since u is a sum of a finite number of normal words which all are consequences of u, we can assume without any loss of generality that u is a normal word itself. Note that if we replace the ideal I by its verbal subideal $I_0 = \mathrm{Id}(u, \Delta^n)$, then all the properties are preserved:

$$ I_0 \supseteq \Delta^n, \quad u \in I_0, \quad u_{(i)} \notin I_0. $$

This leads to the following definition.

Definition. *A normal word* $u = u(x_1, \ldots, x_m) = \sum_{i=0}^{\infty} u_{(i)}$ *is called a special word of type* (n, m, k) *if it involves* m *indeterminates, the weight[3] of* u *is equal to* k, *and if for some* i

$$ u_{(i)} \notin \mathrm{Id}(u, \Delta^n). $$

It follows from the above that the problem of finding a nonhomogeneous stable variety is equivalent to finding some special word. For example, Theorem 1.5.1 actually states that $u(x_1, x_2) = 2z_1 z_2 z_1 - z_1 z_2^2 z_1$ is a special word of type (5,2,3), since for this word we have $m = 2$, $k = 3$ and $u_{(3)} = 2z_1 z_2 z_1 \notin \mathrm{Id}(u, \Delta^5)$. The essence of the following theorem is that *there are no special words of smaller types.* More exactly, note that the set of all triples (n, m, k) is totally ordered under the left lexicographic ordering. Denote this ordering by $<$.

1.5.12. Theorem (Vovsi [99]). *Let* $\mathrm{char}\, K \neq 2, 3$. *If* $(n, m, k) \lneqq (5, 2, 3)$, *then there are no special words of type* (n, m, k).

Sketch of Proof. First, since all 4-stable varieties are homogeneous (Theorem 1.4.2), for any triple of the form $(4, m, k)$ there are no special words of the corresponding type. Next, it is easy to show that there are no special words of type $(n, 1, k)$, where n and k are arbitrary positive integers. Take an arbitrary word $u = \sum a_i z^i$ $(a_i \in K)$ of weight k in one variable, then $a_k \neq 0$. Let $\equiv$ denote equality modulo $I = \mathrm{Id}(u, \Delta^n)$, then

$$ u \equiv a_k z^k + a_{k+1} z^{k+1} + \cdots + a_{n-1} z^{n-1} $$
$$ \equiv z^k (a_k + a_{k+1} z + \cdots + a_{n-1} z^{n-k-1}) \equiv 0. \tag{17} $$

---

[3] Recall that the *weight* $\mathrm{wt}\, u_k$ of an element $u \in KF$ is the number k such that $u \in \Delta^k$ but $u \notin \Delta^{k+1}$.

Denote $a_{k+1}z + \cdots + a_{n-1}z^{n-k-1} = b$. Then $a_k + b$, being a sum of an invertible element and a nilpotent element, is invertible modulo I. By (17), $z^k \equiv 0$. Therefore all the homogeneous components of u belong to I, whence u is not a special word.

It remains to prove that there are no special words of types (5,2,1) and (5,2,2). In the first case it is trivial, since a normal word of weight k with more than k variables simply cannot exist. But the second case is much more difficult and requires several pages of rather tiresome calculations. For the details we refer to [99].

1.6. Stability of unipotent varieties.

A well known theorem of Kolchin [42] states that if G is a matrix group over an arbitrary field and $(g - 1)^n = 0$ for every $g \in G$, then there is a basis in which all matrices from G are simultaneously unitriangular. In terms of varieties or group representations, this theorem means that if a finite-dimensional representation is unipotent, then it is stable. There naturally arises a question whether the requirement on the representation to be finite-dimensional is essential, i.e. *is it true that an arbitrary unipotent representation over a field is stable?* In other words, *does the identity*

$$y \circ (x - 1)^n \equiv 0$$

imply the identity

$$y \circ (x_1 - 1)\dots(x_N - 1) \equiv 0$$

for some $N = N(n)$?

If the ground field has a prime characteristic, the negative answer is immediate. Indeed, let char $K = p$ and let G be an infinite group of exponent p, then it is easy to see that $\text{Reg}_K G$ is unipotent but not stable. However, over a field of characteristic zero this question known as the Unipotency Problem has remained unsolved for a long time. It was discovered, in particular, that it is equivalent to a well known problem on Lie algebras with the Engel condition: *is it true that every Lie algebra over a field of characteristic zero satisfying some Engel identity is nilpotent?*

Recently E. I. Zel'manov [105] has solved the second problem in the affirmative. As a result, the Unipotency Problem has been also solved in the affirmative.

1.6.1. Theorem. *A unipotent representation over a field of characteristic zero is stable.*

In the present section this statement will be proved "modulo" Zel'manov's theorem. The reader interested in the proof of the latter is referred to the original paper [105] or to Chapter 6 of [43].

The transition from Engel Lie algebras to unipotent group representations is well known and can be realized in several ways. We will follow the approach of Heineken [30]. Thus, let $\rho = (V, G)$ be a representation over a field of characteristic 0 satisfying the identity $(x - 1)^n$. Our aim is to prove that ρ is stable. Without loss of generality, we may assume that ρ is faithful and $G \subset \operatorname{End}_K V$. As usual, $(\operatorname{End}_K V)^-$ denotes the Lie algebra on the vector space $\operatorname{End}_K V$ with respect to the operation $[x, y] = xy - yx$.

For any $g \in G$ consider in $\operatorname{End}_K V$ the element

$$\log g = (g - 1) - \frac{(g-1)^2}{2} + \frac{(g-1)^3}{3} - \dots.$$

By assumption, $(g - 1)^n = 0$, therefore this sum is finite and $(\log g)^n = 0$. On the other hand, for an arbitrary nilpotent element a from $\operatorname{End}_K V$ (for example, $a = \log g$) let

$$\exp a = 1 + a + \frac{a^2}{2} + \dots.$$

A straightforward verification shows that

$$\exp(\log g) = g \tag{1}$$

and that for an arbitrary integer r

$$\log(g^r) = r \log g. \tag{2}$$

Denote by L the subset of $\operatorname{End}_K V$ consisting of all linear combinations of the elements $\log g$, $g \in G$, with rational coefficients. The following three lemmas were proved in [30] (the last of them is actually a version of a result of Higgins [32]).

1.6.2. Lemma. $l^n = 0$ *for every* $l \in L$.

Proof. We have to show that if $a_i = \log g_i$ for some $g_1, \ldots, g_m \in G$ and r_i are rational numbers, then

$$(r_1 a_1 + \cdots + r_m a_m)^n = 0.$$

Multiplying the r_i by their common denominator, we may assume that all the r_i are integers. By (1) and (2),

$$g_i = \exp a_i, \qquad g_i^{r_i} = \exp(r_i a_i) = \sum_{k=0}^{n-1} \frac{(r_i a_i)^k}{k!},$$

whence

$$g = g_1^{r_1} \ldots g_m^{r_m} = (1 + r_1 a_1 + \ldots) \ldots (1 + r_m a_m + \ldots)$$
$$= 1 + A_1 + A_2 + \cdots + A_{(n-1)m}$$

where A_i is a homogeneous expression of degree i in $a_1, \ldots, a_m$. Note that $A_1 = r_1 a_1 + \cdots + r_m a_m$.

Since the representation $\rho = (V, G)$ is n-unipotent, it follows that

$$(A_1 + A_2 + \cdots + A_{(n-1)m})^n = (g - 1)^n = 0.$$

Expanding the left-hand side, one can rewrite this equality in the form

$$B_n + B_{n+1} + \cdots + B_t = 0 \qquad\qquad (3)$$

where $t = (n - 1)mn$ and B_i is a homogeneous expression of degree i in $a_1, \ldots, a_m$. In particular, $B_n = A_1^n$. The equality (3) remains valid if we multiply all the a_i by some positive integer k (that is, if we replace each g_i by g_i^k).

Thus there arises the following system of linear equalities in "indeterminates" B_i:

$$k^n B_n + k^{n+1} B_{n+1} + \cdots + k^t B_t = 0 \qquad (k = 1, \ldots, t - n + 1).$$

The determinant of this system is actually the well known Vandermonde determinant; it is nonzero and therefore all the B_i must be equal to zero. In particular, $B_n = A_1^n = (r_1 a_1 + \cdots + r_m a_m)^n = 0$, as required. $\square$

1.6.3. Lemma. L is a rational Lie subalgebra of $(\mathrm{End}_K V)^-$.

Proof. It suffices to show that if $a = \log g$ and $b = \log h$ for some $g, h \in G$, then $[a, b] \in L$. By (1) and (2),

$$g = \exp a = \sum_{i=0}^{n-1} \frac{a^i}{i!}, \qquad h = \exp b = \sum_{j=0}^{n-1} \frac{b^j}{j!}.$$

Therefore

$$
\begin{aligned}
\log(gh - 1) &= \sum_{k=1}^{n-1} \frac{(-1)^{k+1}}{k}(gh - 1)^k \\
&= \sum_{k=1}^{n-1} \frac{(-1)^{k+1}}{k}\left[\left(\sum_{i=0}^{n-1} \frac{a^i}{i!}\right)\left(\sum_{j=0}^{n-1} \frac{b^j}{j!}\right) - 1\right]^k \qquad (4) \\
&= \sum_{p,q=0}^{(n-1)^2} A_{pq}
\end{aligned}
$$

where the expression A_{pq} is homogeneous of degree p in a and homogeneous of degree q in b. Now let r and s be arbitrary positive integers. Replacing in (4) g by g^r and h by h^s (or, equivalently, multiplying a by r and h by s), we obtain the system of inequalities

$$\sum_{p,q=0}^{(n-1)^2} r^p s^q A_{pq} = \log(g^r h^s) \qquad (r, s \in \mathbb{N})$$

in the "indeterminates" A_{pq}. Since r and s are arbitrary, the same argument with the Vandermonde determinant shows that all the A_{pq} can be expressed as linear combinations of the elements $\log(g^r h^s)$ with rational coefficients (by the Cramer formulas), i.e. all the A_{pq} are contained in L. In particular, $A_{11} \in L$. However, expanding (4), it is easy to calculate that

$$A_{11} = ab - \frac{1}{2}(ab + ba) = \frac{1}{2}[a, b].$$

Consequently, $[a, b] = 2A_{11} \in L$. $\square$

1.6.4. Lemma. *Let A be an associative algebra over a field of characteristic zero and let L be a Lie subalgebra of A^- such that A is generated by L. If L is soluble of class t and $l^n = 0$ for every $l \in L$, then A is nilpotent of class n^t.*

P r o o f. a) We proceed by induction on t. Let $t = 1$, that is $[l_1, l_2] = l_1 l_2 - l_2 l_1 = 0$ for all $l_1, l_2 \in L$. By the condition, for arbitrary $l_1, \ldots, l_n \in L$ and arbitrary integers $k_1, \ldots, k_n$

$$(k_1 l_1 + \cdots + k_n l_n)^n = 0.$$

Expanding the left-hand side of this equality and taking into account that the elements l_i commute, we obtain

$$\sum_{m_1 + \cdots + m_n = n} k_1^{m_1} \ldots k_n^{m_n} A_{m_1 \ldots m_n} = 0 \tag{5}$$

where

$$A_{m_1 \ldots m_n} = \frac{n!}{m_1! \ldots m_n!} l_1^{m_1} \ldots l_n^{m_n}.$$

Since (5) is valid for arbitrary integers $k_1, \ldots, k_n$, we again can apply the standard argument with the Vandermonde determinant (more precisely, we should apply it n times, as in the proof of Lemma 1.2.1). Eventually we obtain that all the $A_{m_1 \ldots m_n}$ are equal to 0. In particular,

$$A_{11 \ldots 1} = (n!) l_1 \ldots l_n = 0.$$

Thus $l_1 \ldots l_n = 0$ for all $l_i \in L$, and since A is generated by L, it follows that A is nilpotent of class n.

b) Let now $t > 1$ and assume that for $t - 1$ the lemma is valid. Since $L^{(t-1)}$, the $(t-1)$-th derived subalgebra of L, is soluble of class 1, it follows from the above that

$$\forall l_i \in L^{(t-1)} : \quad l_1 \ldots l_n = 0. \tag{6}$$

Denote by I the ideal of A generated by $L^{(t-1)}$ and let $\bar{A} = A/I$, $\bar{L} = (L + I)/I$. Then $\bar{A}$ and $\bar{L}$ satisfy all conditions of the lemma, but $\bar{L}$ is soluble of class $t - 1$. By the induction hypothesis $\bar{A}$ is nilpotent of class $n^{t-1} = s$, that is

$$\forall a_i \in A : \quad a_1 \ldots a_s \in I.$$

If we prove that

$$\forall u_i \in I : \quad u_1 \ldots u_n = 0, \tag{7}$$

it will be established that A is nilpotent of class $sn = n^t$, as required.

c) Thus, let $u_1, \ldots, u_n \in I$. Since A is generated by L, each u_i is a sum of elements of the form

$$a_1 \ldots a_m l a_{m+1} \ldots a_r \qquad (l \in L^{(t-1)}, \ a_i \in L).$$

Hence to prove (7) it is enough to show that

$$a_{01} \ldots a_{0m_0} l_1 a_{11} \ldots a_{1m_1} l_2 a_{21} \ldots l_n a_{n1} \ldots a_{nm_n} = 0 \tag{8}$$

for arbitrary $l_i \in L^{(t-1)}$ and $a_{ij} \in L$. First of all, it is easy to see that (8) can be reduced to equations of the same type in which the number of factors a_{1i} between l_1 and l_2 equals $m_1 - 1$. Indeed, $l_1 a_{11} = [l_1, a_{11}] + a_{11} l_1$ and since $l' = [l_1, a_{11}] \in L^{(t-1)}$, one can rewrite (8) as

$$a_{01} \ldots a_{0m_0} l' \underbrace{a_{12} \ldots a_{1m_1}}_{m_1-1} l_2 a_{21} \cdots + a_{01} \ldots a_{0m_0} a_{11} l_1 \underbrace{a_{12} \ldots a_{1m_1}}_{m_1-1} l_2 a_{21} \cdots = 0$$

where both terms are of desired form. Repeating this argument, one can eliminate all factors between l_1 and l_2. In other words, instead of (8) it suffices to prove that

$$a_{01} \ldots a_{0m_0} l_1 l_2 a_{21} \ldots a_{2m_2} l_3 a_{31} \ldots l_n a_{n1} \ldots a_{nm_n} = 0.$$

Applying the same argument successively to $a_{21}, \ldots, a_{2m_2}, a_{31}, \ldots$, we eventually reduce (8) to the following assertion:

$$a_1 \ldots a_m l_1 l_2 \ldots l_n a_{m+1} \ldots a_r = 0$$

for all $l_i \in L^{t-1)}$ and $a_i \in L$. But this follows directly from (6). $\square$

Proof of Theorem 1.6.1. Note first that the Lie algebra L satisfies the Engel identity of class $2n$, that is, the identity

$$[\dots[[x,\underbrace{y,]y,],\dots,y]}_{2n} = 0. \tag{5}$$

Indeed,

$$[x,y] = xy - yx,$$

$$[[x,y],y] = xyy - yxy - yxy + yyx,$$

$$\dots\dots\dots\dots$$

$$[\dots[[x,\underbrace{y,]y,],\dots,y]}_{2n} = \sum_{p+q=2n} y^p x y^q.$$

For each summand of the last sum either $p \geq n$ or $q \geq n$, therefore by Lemma 1.6.2 all these summands are equal to zero. Thus L satisfies the identity (5). By [105], L is nilpotent.

Denote by A the rational associative subalgebra of $\mathrm{End}_K V$ generated by all $g - 1$, $g \in G$. Then $L \subset A$ and since

$$g - 1 = \exp(\log g) - 1 = \log g + \frac{\log^2 g}{2!} + \dots + \frac{\log^{n-1} g}{(n-1)!},$$

A is generated by L as an associative algebra. By Lemmas 1.6.2–4, A is nilpotent of some class N. In particular, $(g_1 - 1)(g_2 - 1)\dots(g_N - 1) = 0$ for arbitrary $g_i \in G$, whence $\rho = (V, G)$ is stable of class N. □

It was noted in the beginning of this section that the Unipotency Problem was, in fact, equivalent to the problem on Lie algebras with the Engel identity. Therefore there arises a natural desire to find a proof of Theorem 1.6.1 not depending on Zel'manov's theorem, thus giving an alternative proof of the latter. In the remainder of the section we will sketch one idea which, had it been successful, would have given such a proof immediately.

Consider the variety $\mathcal{U}_n = [z^n]$ of n-unipotent representations of groups over a field K of characteristic 0. In view of the extreme simplicity of the identity z^n, the following question suggests itself: is $\mathcal{U}_n$ homogeneous? Suppose the answer is "yes". Since $\mathcal{U}_n$ is also a Magnus variety [51], it follows from Theorem 1.3.1 that there exists a variety $\mathcal{M}$ of associative algebras such that $\mathcal{M}^\alpha = \mathcal{U}_n$. From the definition of the map α it is clear that $\mathcal{M}$ must satisfy the identity z^n whence, by the well known theorem of Nagata [66] (see also [35]), $\mathcal{M}$ is nilpotent. Applying 1.3.1 again, we obtain that $\mathcal{U}_n$ is stable, as desired!

Alas, our naive hope to solve the problem so easily fails:

1.6.4. Proposition (Vovsi [99]). *The variety $\mathcal{U}_3$ is not homogeneous.*

P r o o f. Denote $U_3 = \operatorname{Id} \mathcal{U}_3$ and suppose that U_3 is a homogeneous verbal ideal. Since $U_3 = \operatorname{Id}(z^3)$, it follows from Lemma 1.2.2 that $\operatorname{lin} z^3 \in U_3$. The representation $\rho = (V, G)$ constructed in § 1.5 belongs to the variety $\mathcal{X}_0$ and, consequently, it satisfies the identity z^3 (see Lemma 1.5.2). Therefore it satisfies the identity

$$\operatorname{lin} z^3 = \sum_{\sigma \in S_3} z_{\sigma(1)} z_{\sigma(2)} z_{\sigma(3)} = v(x_1, x_2, x_3)$$

as well. On the other hand, using the notation and the formulas (9) and (11) from § 1.5, we get

$$
\begin{aligned}
v(a_1, a_2, a_1) &= 2(\bar{a}_1^2 \bar{a}_2 + \bar{a}_1 \bar{a}_2 \bar{a}_1 + \bar{a}_2 \bar{a}_1^2) \\
&= 2\left(e_{19} - e_{39} - e_{30} - \frac{1}{2}e_{19} - \frac{1}{2}e_{10} + e_{10} + e_{39} + e_{30}\right) \\
&= e_{19} + e_{10} \neq 0,
\end{aligned}
$$

so that $v(x_1, x_2, x_3)$ is not an identity of ρ. This contradiction shows that $\mathcal{U}_3$ cannot be homogeneous. $\square$

One particular consequence of this result is that, apart from the variety $\mathcal{X}_0$ constructed in the previous section, we now have another example of a stable nonhomogeneous variety of group representations, namely the variety $\mathcal{U}_3$. This example is certainly nicer, but it should be noted that our proof of Proposition 1.6.4 is essentially based on the properties of $\mathcal{X}_0$.

Chapter 2

LOCALLY FINITE-DIMENSIONAL VARIETIES

One of the most interesting classes of varieties of arbitrary algebraic structures is the class of locally finite varieties. This general principle applies, in particular, in the theory of varieties of group representations. However, in the case of group representations, as opposed to more "usual" algebraic structures, it is natural to investigate *two* closely related notions: locally finite-dimensional varieties and, as a special case, locally finite varieties. The present chapter deals with these two classes of varieties.

Throughout the chapter the ground ring K is a field. In § 2.1 we establish a few basic properties of locally finite and locally finite-dimensional varieties. The main results of § 2.2 give a nice and somewhat unexpected characterization of such varieties in terms of stable varieties of representations and locally finite varieties of groups. This characterization is based on a result of Gringlaz [25] and Theorem 1.6.1.

Locally finite and locally finite-dimensional varieties have important numerical invariants—the so-called order functions—which are studied in § 2.3. The last two sections, §§ 2.4 and 2.5, are devoted to critical representations. The material of these sections is more or less standard; however, it has applications specific for the context of group representations. For example, it implies that over a field of zero characteristic the lattice of subvarieties of every locally finite variety is distributive (Corollary 2.5.5).

But the major applications of the technique of critical representations will be presented in Chapter 3. In that chapter we continue to study locally finite and locally finite-dimensional varieties, keeping in mind one major theme: the finite basis problem.

2.1. Basic notions and properties

We start with a few simple definitions. A representation $\rho = (V, G)$ is called *finitely generated* if G is a finitely generated group and V is a finitely generated KG-module. A representation $\rho = (V, G)$ is called *finite* if it is finite-dimensional and the group $\bar{G} = G/\mathrm{Ker}\,\rho$ is finite; in this case the number $\dim V + |\bar{G}|$ is called the *order* of the representation ρ and is denoted by $|\rho| = |(V, G)|$. Finally, a representation is said to be *locally finite-dimensional* or *locally finite* if all its finitely generated subrepresentations are respectively finite-dimensional or finite.

2.1.1. Lemma. *Let $\rho = (V, G)$ be a representation, N a normal subgroup of G, and let (V, N) be the restriction of the representation ρ to N. If (V, N) is locally finite-dimensional (locally finite) and the group G/N is locally finite, then ρ is locally finite-dimensional (locally finite) as well.*

P r o o f. We consider only the case of finite-dimensional (V, N); the second case is similar. Let $\sigma = (W, H)$ be a finitely generated subrepresentation of ρ. Take its subrepresentation $(W, N \cap H)$ and the group $H/(N \cap H)$. The latter, being a finitely generated subgroup of a locally finite group G/N, is finite. By the Schreier formula, $N \cap H$ is finitely generated as well. Further, since W is a finitely generated KH-module, we have

$$W = w_1 \circ KH + \cdots + w_n \circ KH$$

for some $w_1, \ldots, w_n$. If now $t_1, \ldots, t_m$ is a complete system of representatives of $(N \cap H)$-cosets in H, then

$$W = \sum_{i=1}^{n} \sum_{j=1}^{m} (w_i \circ t_j) \circ K[N \cap H].$$

Therefore $(W, N \cap H)$ is a finitely generated subrepresentation of (V, N). Since the latter is locally finite-dimensional, W is finite-dimensional, as desired. □

2.1.2. Lemma. *Let $\rho = (V, G)$ be a representation and let A be a G-submodule of V such that the corresponding representations (A, G) and*

$(V/A, G)$ are locally finite-dimensional. Then ρ is locally finite-dimensional as well.

P r o o f. It suffices to prove that if $H = \langle h_1, \ldots h_r \rangle$ is a finitely generated subgroup of G and $v \in V$, then $\dim_K(v \circ KH) < \infty$. Denote by $H^{(t)}$ the set of all elements of the form $h_{i_1}^{\pm 1} h_{i_2}^{\pm 1} \ldots h_{i_s}^{\pm 1}$ where $1 \leq i_k \leq r$ and $s \leq t$, and let $KH^{(t)}$ be the *set* of all K-linear combinations of elements from $H^{(t)}$. All the sets $H^{(t)}$ are finite and $H = \bigcup_{t=1}^{\infty} H^{(t)}$. Therefore the desired assertion is equivalent to the following: there exists t such that

$$v \circ KH^{(t)} = v \circ KH^{(t+1)} = \cdots = v \circ KH. \tag{1}$$

Consider the quotient representation $(V/A, G)$. Since it is locally finite-dimensional, we see that for the element $\bar{v} = v + A$ there exists n such that

$$\bar{v} \circ KH^{(n)} = \bar{v} \circ KH^{(n+1)} = \cdots = \bar{v} \circ KH. \tag{2}$$

Let $g_1, \ldots, g_k$ be all elements of the (finite) set $H^{(n+1)}$. It follows from (2) that for every $i = 1, \ldots, k$

$$v \circ g_i = a_i + w_i, \quad \text{where} \quad a_i \in A, w_i \in v \circ KH^{(n)}.$$

Then

$$a_i = v \circ g_i - w_i \in v \circ KH^{(n+1)}. \tag{3}$$

Further, since (A, G) is a locally finite-dimensional representation, for every $i = 1, \ldots, k$ there exists m_i such that

$$a_i \circ KH^{(m_i)} = a_i \circ KH^{(m_i+1)} = \ldots . \tag{4}$$

Now, if $m = \max_{1 \leq i \leq k} m_i$ and $h \in H^{(m+1)}$, then by (4) and (3) we have

$$a_i \circ h \in a_i \circ KH^{(m+1)} = a_i \circ KH^{(m)} \subseteq v \circ KH^{(m+n+1)},$$

whence $v \circ g_i h = a_i \circ h + w_i \circ h \in v \circ KH^{(m+n+1)}$. But elements of the form $g_i h$ run over all the set $H^{(m+n+2)}$. Therefore if we set $t = m + n + 1$, then $v \circ KH^{(t)} = v \circ KH^{(t+1)} = \cdots = v \circ KH$. $\square$

A variety of group representations is called *locally finite-dimensional* (*locally finite*) if all its representations are locally finite-dimensional (locally finite).

Examples. 1. Let Θ be a locally finite variety of groups. It is easy to see that $\omega\Theta$ is a locally finite variety of group representations.

2. Every stable variety is locally finite-dimensional. Indeed, if $\rho = (V, G)$ is a stable representation, then V possesses a series of G-submodules

$$0 = A_0 \subset A_1 \subset \cdots \subset A_n = V$$

such that G acts trivially on each quotient A_{i+1}/A_i. The corresponding representations $(A_{i+1}/A_i, G)$ are, of course, locally finite-dimensional. By Lemma 2.1.2, ρ is also locally finite-dimensional.

The collection of examples of locally finite and locally finite-dimensional varieties can be substantially extended, if we use the following immediate corollary of Lemmas 2.1.1 and 2.1.2.

2.1.3. Proposition. (i) *If varieties $\mathcal{X}$ and $\mathcal{Y}$ are locally finite-dimensional, then the variety $\mathcal{X}\mathcal{Y}$ is also locally finite-dimensional.*

(ii) *If $\mathcal{X}$ is a locally finite-dimensional (locally finite) variety of representations , Θ a locally finite variety of groups, then the variety $\mathcal{X} \times \Theta$ is locally finite-dimensional (locally finite).* $\square$

It follows from Proposition 2.1.3 and the above examples that if Θ is a locally finite variety of groups, then every subvariety of $S^n \times \Theta$ is locally finite-dimensional, but every subvariety of $S \times \Theta = \omega\Theta$ is locally finite. In the next section we will prove that *this is actually a complete characterization of locally finite-dimensional and locally finite varieties of group representations.*

We establish now a few simple properties of locally finite-dimensional varieties. Recall that an algebra over a field is said to be locally finite-dimensional if each of its finitely generated subalgebras is finite-dimensional.

2.1.4. Proposition. *A variety $\mathcal{X}$ is locally finite-dimensional if and only if the algebra $KF/\mathrm{Id}\,\mathcal{X}$ is locally finite-dimensional.*

Proof. Let $\mathcal{X}$ be locally finite-dimensional, $I = \text{Id}\,\mathcal{X}$. Take an arbitrary finite subset $\bar{u}_i = \{u_i + I | i = 1, \ldots, n\}$ of KF/I and show that the subalgebra generated by this subset is finite-dimensional. Let

$$u_i = \sum \lambda_{ij} f_{ij}, \quad \text{where} \quad \lambda_{ij} \in K, f_{ij} \in F.$$

Denote by G the subgroup of F generated by all f_{ij}, and by V the cyclic G-submodule of KF/I generated by the element $\bar{1} = 1 + I$. Consider the corresponding subrepresentation $\rho = (V, G)$ of $\text{Fr}\,\mathcal{X} = (KF/I, F)$. Since ρ is finitely generated, we see that $\dim V < \infty$. It is clear that V is actually a subalgebra containing the elements $\bar{u}_1, \ldots, \bar{u}_n$. Therefore these elements generate a finite-dimensional subalgebra.

Conversely, suppose that KF/I is a locally finite-dimensional algebra, and let $\rho = (V, G)$ be a finitely generated representation from $\mathcal{X}$. We show that $\dim V < \infty$. Without loss of generality, one may assume that V is a cyclic KG-module. Therefore if G is a group with n generators, then ρ is an epimorphic image of the representation $\text{Fr}_n\mathcal{X} = (KF_n/I_n, F_n)$. By 0.3.4, $(KF_n/I_n, F_n) \subset (KF/I, F)$. Since the algebra KF_n/I_n is finitely generated, $\dim(KF_n/I_n) < \infty$ and, consequently, $\dim V < \infty$. $\square$

2.1.5. Lemma. *Let $\mathcal{D}$ be a class of representations. If the variety $\mathcal{X} = \text{var}\,\mathcal{D}$ is locally finite-dimensional, then all finitely generated representations from $\mathcal{X}$ belong to the class* $\text{VQSD}_0\,\mathcal{D}$.

Proof. Let $\rho = (V, G)$ be a finitely generated representation from $\mathcal{X}$ and let $\bar{\rho} = (V, \bar{G})$ be its faithful image. Then $\bar{\rho} \in \text{QSC}\,\mathcal{D}$, so that there exists $\sigma = (W, H) \in \text{SC}\mathcal{D}$ with an epimorphism $\sigma \to \bar{\rho}$. The representation σ can be also chosen to be finitely generated, and therefore $\dim W < \infty$.

By the choice of $\sigma = (W, H)$, there exist representations $\tau_i \in \mathcal{D}$, $i \in I$, such that $\sigma \subseteq \overline{\prod}_{i \in I} \tau_i = \tau$. Since W is finite-dimensional, there is a finite subset I_0 of I such that the natural projection $\pi : \tau \to \overline{\prod}_{i \in I_0} \tau_i$ is injective on W. Therefore the restriction $\pi : \sigma \to \sigma^\pi$ is a left epimorphism of the representation $\sigma = (W, H)$ onto σ^π. Since $\sigma^\pi \in \text{SD}_0\mathcal{D}$, it follows that $\sigma \in \text{VSD}_0\mathcal{D}$, whence $\bar{\rho} \in \text{QVSD}_0\,\mathcal{D}$ and $\rho \in \text{VQVSD}_0\,\mathcal{D}$. Using an easily verified relation $\text{QV} \le \text{VQ}$ between the closure operations V and Q, we obtain $\rho \in \text{VQSD}_0\,\mathcal{D}$. $\square$

2.1.6. Proposition. *If $\mathcal{X}$ is a locally finite-dimensional variety, then the class of all locally finite representations from $\mathcal{X}$ is a subvariety.*

Proof. Let $\mathcal{D}$ be the class of all finite representations from $\mathcal{X}$ and $\mathcal{X}_0 = \text{var}\,\mathcal{D}$. It is clear that all locally finite representations from $\mathcal{X}$ belong to $\mathcal{X}_0$. Further, choose a finitely generated representation ρ in $\mathcal{X}_0$. By the above lemma, $\rho \in \text{VQSD}_0\,\mathcal{D}$. Since $\mathcal{D}$ is closed under V, Q, S, D_0, it follows that $\rho \in \mathcal{D}$, so that ρ is finite. Therefore $\mathcal{X}_0$ consists of locally finite representations. □

A representation is called *locally stable* if all its finitely generated subrepresentations are stable. For example, an arbitrary representation of a locally finite p-group over a field of characteristic p is locally stable (this follows, e.g., from [5]). According to Example 2 above, locally stable representations are locally finite-dimensional.

2.1.7. Proposition. *If $\mathcal{X}$ is a locally finite-dimensional variety, then the class of all locally stable representations from $\mathcal{X}$ is a subvariety.*

Proof. Denote by $\mathcal{D}$ the class of all stable representations from $\mathcal{X}$ and repeat the proof of the previous assertion. □

2.2. Characterization of locally finite and locally finite-dimensional varieties

The purpose of the present section is to prove the following theorems.

2.2.1. Theorem. *A variety $\mathcal{X}$ of group representations is locally finite-dimensional if and only if $\mathcal{X} \subseteq S^n \times \Theta$ for some n and some locally finite variety of groups Θ.*

2.2.2. Theorem. *A variety $\mathcal{X}$ of group representations is locally finite if and only if $\mathcal{X} \subseteq S \times \Theta = \omega\Theta$ for some locally finite variety of groups Θ.*

These theorems have the following history. In 1971, Gringlaz [25] proved that every locally finite-dimensional variety is contained in some

$\mathcal{X}' \times \Theta$, where $\mathcal{X}'$ is a locally stable variety of representations, Θ a locally finite variety of groups. In particular, this implied Theorem 2.2.1 in prime characteristic and Theorem 2.2.2. But fifteen years later it became clear that over a field of characteristic zero every locally stable variety is actually stable (indeed, a locally stable variety is certainly unipotent, and it remains to apply Theorem 1.6.1). Thus Theorem 2.2.1 was completed.

In proving Theorems 2.2.1 and 2.2.2 we follow [25]. First, let us prove several auxiliary facts.

2.2.3. Lemma. *If $\mathcal{X}$ is a locally finite-dimensional variety, then there exists a polynomial $\varphi(t)$ such that for any representation (V, G) from $\mathcal{X}$ the eigenvalues of any $g \in G$ are roots of this polynomial.*

P r o o f. Let $\mathrm{Fr}_1 X = (KF_1/I_1, F_1)$ be a free cyclic representation of rank 1 of the variety $\mathcal{X}$ (here F_1 is an infinite cyclic group with a generator x). Since $\mathcal{X}$ is locally finite-dimensional, the space KF_1/I_1 is finite-dimensional. Regarding x as a linear operator of this space, denote by $\varphi(t)$ its minimal annihilating polynomial. Further, for any $g \in G$ let λ be an eigenvalue of g, and let $v \in V$ be a corresponding eigenvector. Then the cyclic representation $\rho = (v \circ K\langle g \rangle, \langle g \rangle)$ is finite-dimensional, and if $f(t)$ is the minimal polynomial of the restriction of the operator g to the subspace $v \circ K\langle g \rangle$, then $f(\lambda) = 0$. We show that $f(t)$ divides $\varphi(t)$. To do this, it suffices to show that $\varphi(g)$ annihilates the space $v \circ K\langle g \rangle$.

Since $\mathrm{Fr}_1 \mathcal{X} = (KF_1/I_1, F_1)$ is a free representation of $\mathcal{X}$ with free generators x and $\bar{1} = 1 + I_1$, the map $\bar{1} \mapsto v$, $x \mapsto g$ extends to an epimorphism of representations $\mu : (KF_1/I_1, F_1) \to (v \circ K\langle g \rangle, \langle g \rangle)$. But $\varphi(x)$ annihilates KF_1/F_1, therefore $\varphi(g)$ annihilates $v \circ K\langle g \rangle$, as required. It follows that $f \mid \varphi$, whence $\varphi(\lambda) = 0$. $\square$

2.2.4. Lemma. *If $\mathcal{X}$ is a locally finite-dimensional variety, then the variety of groups generated by all finite groups admitting a faithful representation in $\mathcal{X}$ is locally finite.*

P r o o f. Denote by Θ the class of all groups satisfying the above condition. Since $\mathrm{var}\,\Theta = QSC\Theta$, it suffices to show that $C\Theta$ consists of locally finite groups.

Let $G_i \in \Theta (i \in I)$, $G = \overline{\prod}_{i \in I} G_i$ and let H be a finitely generated subgroup of G. We have to prove that H is finite. For each $i \in I$ pick in $\mathcal{X}$ a faithful representation $\rho_i = (V_i, G_i)$ of the group G_i. Let $n_i = |G_i|$ and let $g_i^{(1)}, \dots, g_i^{(n_i - 1)}$ be all nonunit elements of G_i. For arbitrary $j = 1, \dots, n_i - 1$ take an isomorphic copy $V_i^{(j)}$ of the G_i-module V_i and let

$$W_i = V_i^{(1)} \oplus V_i^{(2)} \oplus \cdots \oplus V_i^{(n_i - 1)}.$$

Denote the corresponding representation (W_i, G_i) by σ_i. Since the representation ρ_i is faithful, for each $g_i^{(j)}$ there exists $v_i^{(j)} \in V_i^{(j)}$ such that $v_i^{(j)} \circ g_i^{(j)} \neq v_i^{(j)}$. If now $w_i = v_i^{(1)} + v_i^{(2)} + \cdots + v_i^{(n_i - 1)} \in W_i$, then

$$w_i \circ g \neq w_i$$

for every $1 \neq g \in G_i$.

Consider the Cartesian product

$$\sigma = \overline{\prod}_{i \in I} \sigma_i = \overline{\prod}_{i \in I} (W_i, G_i) = (W, G)$$

and choose in W the element $w = (w_i)$ whose components are the w_i. Evidently, $w \circ g \neq w$ for every $1 \neq g \in G$. Therefore the cyclic subrepresentation $\rho = (w \circ KH, H) = (A, H)$ of σ is faithful. Let π_i be the natural projection of σ onto σ_i. It induces a homomorphism of $\rho = (A, H)$ into σ_i. If (A_i, H_i) is the kernel of this homomorphism, then

$$\rho^{\pi_i} \simeq (A/A_i, H/H_i)$$

and so the group H/H_i is finite. Clearly $\bigcap_{i \in I} A_i = 0$.

Since ρ is a finitely generated representation of a locally finite-dimensional variety $\mathcal{X}$, its domain A is finite-dimensional. Therefore there exists a finite subset I_0 of I such that $\bigcap_{i \in I_0} A_i = 0$. For each $i \in I$ the subgroup H_i acts trivially on A/A_i, therefore $H^* = \bigcap_{i \in I_0} H_i$ acts trivially on every A/A_i, $i \in I_0$ and hence on $A/\bigcap_{i \in I_0} A_i = A$. Since ρ is faithful, it follows that $H^* = 1$ whence

$$H \simeq H/H^* \hookrightarrow \prod_{i \in I_0} H/H_i$$

and the group H is finite. $\square$

Now let K be a subfield of a field L. According to §0.4, we have the natural maps

$$\nu : \mathrm{M}(K) \to \mathrm{M}(L) \ \text{ and } \ \nu' : \mathrm{M}(L) \to \mathrm{M}(K). \qquad (1)$$

The next two lemmas deal with these maps.

2.2.5. Lemma. *If a variety $\mathcal{X} \in \mathrm{M}(K)$ is locally finite-dimensional, then the variety $\mathcal{X}^\nu \in \mathrm{M}(L)$ is also locally finite-dimensional.*

Proof. Let $\rho = (V, G) \in \mathcal{X}^\nu$. It is enough to prove that if H is a finitely generated subgroup of G and $v \in V$, then $\dim_L(v \circ LG) < \infty$. We are given that ρ, regarded as a representation over K, is locally finite-dimensional. Therefore $\dim_K(v \circ KG) < \infty$, whence the required inequality follows. $\square$

2.2.6. Lemma. *If a variety $\mathcal{Y} \in \mathrm{M}(L)$ is locally stable, then the variety $\mathcal{Y}^{\nu'} \in \mathrm{M}(K)$ is also locally stable.*

Proof. It suffices to prove that for arbitrary n the free cyclic representation $\mathrm{Fr}_n(\mathcal{Y}^{\nu'})$ of rank n in $\mathcal{Y}^{\nu'}$ is stable. Consider first the representation $\mathrm{Fr}_n \mathcal{Y}$. It has the form

$$\mathrm{Fr}_n \mathcal{Y} = (LF_n/J_n, F_n),$$

where $J_n = \mathrm{Id}_n \mathcal{Y}$ is the ideal of identities of $\mathcal{Y}$ in the algebra LF_n. By assumption, this representation is stable. Since $\mathrm{Id}_n(\mathcal{Y}^{\nu'}) = J_n \cap KF_n$, we have

$$\mathrm{Fr}_n(\mathcal{Y}^{\nu'}) = (KF_n/(J_n \cap KF_n), F_n),$$

and it is clear that $KF_n/(J_n \cap KF_n)$ is a KF_n-submodule of LF_n/J_n. Therefore $\mathrm{Fr}_n(\mathcal{Y}^{\nu'})$ is also stable. $\square$

2.2.7. Proposition (Gringlaz [25]). *A variety of group representations $\mathcal{X}$ is locally finite-dimensional if and only if $\mathcal{X} \subseteq \mathcal{X}' \times \Theta$, where $\mathcal{X}'$ is* *a locally stable variety of representations and Θ is a locally finite variety of groups.*

P r o o f. Since a locally stable variety is locally finite-dimensional, "if" follows directly from Proposition 2.1.3 (ii). To prove the nontrivial part, note first that we may suppose the ground field K to be algebraically closed. Indeed, if it is not the case, then we embed K in its algebraic closure L and consider the maps (1). Let $\mathcal{X}$ be a locally finite-dimensional variety over K. By Lemma 2.2.5, $\mathcal{X}^\nu$ is a locally finite-dimensional variety over L. Suppose our assertion has been proved over an algebraically closed field. Then $\mathcal{X}^\nu \subseteq \mathcal{Y} \times \Theta$, where $\mathcal{Y}$ is a locally stable variety of representations over L, Θ is a locally finite variety of groups. Using 0.4.2, we have

$$\mathcal{X} = \mathcal{X}^{\nu\nu'} \subseteq (\mathcal{Y} \times \Theta)^{\nu'} \subseteq \mathcal{Y}^{\nu'} \times \Theta \,,$$

where $\mathcal{Y}^{\nu'}$ is locally stable by Lemma 2.2.6.

Thus, from now on K is algebraically closed. Let $\mathcal{X}$ be a locally finite-dimensional variety of group representations over K, $\mathcal{X}'$ the class of all locally stable representations from $\mathcal{X}$, and Θ the locally finite variety of groups defined in Lemma 2.2.4. Since $\mathcal{X}'$ is a variety in view of Lemma 2.1.7, it remains to prove that $\mathcal{X} \subseteq \mathcal{X}' \times \Theta$. It suffices to show that every finitely generated representation $\rho = (V, G)$ from $\mathcal{X}$ belongs to $\mathcal{X}' \times \Theta$. Since $\mathcal{X}$ is locally finite-dimensional, ρ is finite-dimensional. Choose in V a composition series of G-modules

$$0 = A_0 \subset A_1 \subset \cdots \subset A_n = V \qquad (2)$$

and let H_i be the kernel of the naturally arising representation $(A_{i+1}/A_i, G)$, $i = 1, \ldots, n$. Then G/H_i can be considered as an irreducible matrix group over K. Since K is algebraically closed, G/H_i is absolutely irreducible. By 2.2.3, all eigenvalues of all elements from G/H_i are roots of some fixed polynomial. Using a classical Burnside's theorem (see e.g. [87, p. 254]), we conclude that G/H_i is finite. Hence the Θ-verbal subgroup $\Theta^*(G)$ of G is contained in $H_i\,(i = 1, \ldots, n)$ and therefore in $H = \cap H_i$. Consequently, $\Theta^*(G)$ acts trivially on each quotient of (2), hence $(V, \Theta^*(G))$ is stable and it follows that $(V, G) \in \mathcal{X}' \times \Theta$. $\square$

2.2.8. Corollary. *A variety* $\mathcal{X}$ *of group representations over a field of prime characteristic is locally finite-dimensional if and only if* $\mathcal{X} \subseteq \mathcal{S} \times \Theta$ *for some locally finite variety of groups* Θ.

P r o o f. To prove the nontrivial part, let $\mathcal{X}$ be a locally finite-dimensional variety over a field K of characteristic p. By 2.2.7, we have $\mathcal{X} \subseteq \mathcal{X}' \times \Theta_1$, where $\mathcal{X}'$ is locally stable but Θ_1 is a locally finite variety of groups. It is well known that a group admitting a faithful n-stable representation over a field of characteristic p is an $(n-1)$-nilpotent p-group of finite exponent (see e.g. [28] or [75]). Therefore it is easy to see that if (V, G) is a faithful representation from $\mathcal{X}'$, then G belongs to a variety Θ_2 of locally finite p-groups depending on $\mathcal{X}'$ only. Thus $\mathcal{X}' \subseteq \mathcal{S} \times \Theta_2$, whence

$$\mathcal{X} \subseteq (\mathcal{S} \times \Theta_2) \times \Theta_1 = \mathcal{S} \times (\Theta_2\Theta_1)$$

where the variety $\Theta_2\Theta_1 = \Theta$ is locally finite. $\square$

In particular, since any variety of the form $\mathcal{S} \times \Theta$ with a locally finite Θ is locally finite itself, we obtain the following rather unexpected fact.

2.2.9. Corollary. *Over a field of prime characteristic, every locally finite-dimensional variety is locally finite.* $\square$

We can now complete the proofs of the main results of this section.

P r o o f o f T h e o r e m 2.2.1. If char $K = p$, the result follows from 2.2.8. If char $K = 0$, it is enough to note that every locally stable variety is unipotent and then apply 2.2.7 and 1.6.1. $\square$

P r o o f o f T h e o r e m 2.2.2. If char $K = p$, the result follows from 2.2.8 again. Let char $K = 0$ and let $\mathcal{X}$ be a locally finite variety over K. By 2.2.7, $\mathcal{X} \subseteq \mathcal{X}' \times \Theta$, where $\mathcal{X}'$ is locally stable and Θ is locally finite. Take in $\mathcal{X}$ an arbitrary faithful finitely generated representation $\rho = (V, G)$ and let $H = \Theta^*(G)$. Then G/H, being a finitely generated group of Θ, is finite. Therefore H is finitely generated, and it is clear that the representation (V, H) is finitely generated as well. Since this representation lies in $\mathcal{X}'$, it is stable, but since it is faithful, we conclude that H is a nilpotent torsion-free group (see e.g. [28]). But H must be finite because (V, H) is a faithful finitely generated representation of a locally finite variety $\mathcal{X}$. It follows that $H = \{1\}$, whence $G \in \Theta$ and $\rho \in \omega\Theta$. Thus $\mathcal{X} \subseteq \omega\Theta$. $\square$

We conclude this section with a few remarks concerning the cardinalities of the sets of locally finite and locally finite-dimensional varieties of representations over an arbitrary field K. Denote these sets by $\mathrm{M}_{lf}(K)$ and $\mathrm{M}_{lfd}(K)$ respectively. According to a theorem of Ol'shansky [70], there is a continuum of locally finite varieties of groups. Since the map $\Theta \mapsto \omega\Theta$ is injective, it follows that over any K there is at least continuum of locally finite varieties of group representations. Further, by a result of Grinberg [79], the cardinality of the set of subvarieties of S^4 is not less that the cardinality $|K|$ of the ground field. All these subvarieties are stable and therefore locally finite-dimensional. If now char $K = p$, then, by 2.2.9, all locally finite-dimensional varieties are locally finite. Thus $|\mathrm{M}_{lf}(K)| \geq \max(\mathfrak{c}, |K|)$, where $\mathfrak{c}$ is the continuum.

On the other hand, since the group algebra KF is of countable dimension, each completely invariant ideal of KF has a countable basis. It is clear that there are exactly $\max(\mathfrak{c}, |K|)$ countable subsets in KF, whence there are at most $\max(\mathfrak{c}, |K|)$ varieties of group representation over any field K.

Let now char $K = 0$. Every locally finite variety $\mathcal{X}$ is generated by its finite representations. But over a field of characteristic zero every finite representation is completely reducible, hence $\mathcal{X}$ is generated by irreducible representations of finite groups. Since the set of such representations is countable, there are at most continuum of locally finite varieties of representations over K. Combining all these remarks, we obtain:

2.2.10. Proposition. *Let K be a field. If* char $K = p$, *then*

$$|\mathrm{M}_{lf}(K)| = |\mathrm{M}_{lfd}(K)| = \max(\mathfrak{c}, |K|),$$

but if char $K = 0$, *then*

$$|\mathrm{M}_{lf}(K)| = \mathfrak{c} \quad \text{and} \quad |\mathrm{M}_{lfd}(K)| = \max(\mathfrak{c}, |K|). \quad \square$$

2.3. The order functions

Let $\mathcal{X}$ be a locally finite-dimensional variety and let

$$\mathrm{Fr}_n \, \mathcal{X} = (KF_n/I_n, F_n) = (E_n, F_n)$$

be its free cyclic representation of a finite rank n (where $I_n = \mathrm{Id}_n \mathcal{X} = \mathcal{X}^*(KF_n, F_n)$). Then the number $l(n) = \dim E_n$ is finite for each n. Thus, to every locally finite-dimensional variety $\mathcal{X}$ one can naturally assign a function $l = l(n)$ called the *left order function* of $\mathcal{X}$. If $\mathcal{X}$ is locally finite, then the number $r(n) = |F_n/\mathrm{Ker}\,(\mathrm{Fr}_n\mathcal{X})|$ is also finite. Therefore to every locally finite variety $\mathcal{X}$ one can naturally assign another function $r = r(n)$ called the *right order function* of $\mathcal{X}$. Note that always $l(n) \le r(n)$.

With a slight abuse of language, we can now say that locally finite-dimensional varieties are varieties with the left order function, but locally finite varieties are varieties with *both* order functions.

2.3.1. Example. Let Θ be a locally finite variety of groups. We know that the variety of representations $\omega\Theta$ is also locally finite. Let us evaluate the order functions of this variety. To do this, we note that if $F_n(\Theta)$ is the free group of rank n in Θ, then its regular representation $\mathrm{Reg}\, F_n(\Theta) = (K[F_n(\Theta)], F_n(\Theta))$ is the free faithful cyclic representation of rank n of in $\omega\Theta$ (cf. Example on p.12). Hence if l and r are the left and the right order functions of $\omega\Theta$ respectively, then for each n

$$l(n) = r(n) = |F_n(\Theta)|.$$

The order functions are important numerical invariants of a variety, giving essential information about the latter. In the present section we prove, following [93, 95], several results concerning the order functions. The proofs of these results are not difficult but they sometimes involve rather interesting techniques: free ideal rings, classical properties of periodic matrix groups, etc.

The first question we are interested in consists in the following. By 2.1.3 (ii), if $\mathcal{X}$ is a locally finite-dimensional (locally finite) variety of representations and Θ a locally finite variety of groups, then the variety $\mathcal{X} \times \Theta$ is also locally finite-dimensional (locally finite). Is it possible to express explicitly the order functions of $\mathcal{X} \times \Theta$ by means of the order functions l and r of $\mathcal{X}$ and the order function f of Θ (recall that by definition $f(n) = |F_n(\Theta)|$)? Let $l^{\times}$ and $r^{\times}$ be the order functions of $\mathcal{X} \times \Theta$.

2.3.2. Proposition. *If $\mathcal{X}$ is a locally finite-dimensional variety of representations and Θ a locally finite variety of groups, then*

$$l^{\times}(n) = l((n-1)f(n) + 1)f(n). \tag{1}$$

In addition, if $\mathcal{X} \neq \mathcal{E}$ is locally finite, then

$$r^{\times}(n) = r((n-1)f(n) + 1)f(n). \tag{2}$$

P r o o f. Take the absolutely free representation $\rho = (KF_n, F_n)$ and let $H = \Theta^*(F_n)$ and $A = \mathcal{X}^*(KF_n, H)$. Then $(KF_n/A, F_n) = \mathrm{Fr}_n(\mathcal{X} \times \Theta)$. By definition,

$$l^{\times}(n) = \dim(KF_n/A). \tag{3}$$

Let j be the index of H in F_n, then $j = f(n)$. By the Schreier formula, H is a free group of rank

$$m = (n-1)j + 1. \tag{4}$$

If $f_1, \ldots, f_j$ is a complete set of representatives of the H-cosets in F_n, then KF_n can be decomposed in the direct sum of H-submodules $KF_n = f_1 KH \oplus \cdots \oplus f_j KH$. The corresponding representations $(f_i KH, H)$ are all isomorphic to the regular representation $\mathrm{Reg}\, H$. Denoting $B = \mathcal{X}^*(\mathrm{Reg}\, H)$ and keeping in mind that verbals permute with direct sums , we obtain

$$A = \bigoplus_{i=1}^{j} f_i B, \quad KF_n/A = \bigoplus_{i=1}^{j} f_i(KH/B). \tag{5}$$

Since $H = F_m$, we have $\dim(KF_n/A) = l(m)j$. Therefore (1) follows from (3) and (4).

To prove (2), we first note that, by (5),

$$(KF_n/A, F_n) = (f_1 \cdot (KH/B) \oplus \cdots \oplus f_j \cdot (KH/B), F_n). \tag{6}$$

Next, let $G = \text{Ker}\,(KF_n/A, F_n)$. We show that $G \subseteq H$. Each $g \in G$ can be presented as $g = f_i h$ where $h \in H$. Suppose that $g \notin H$, then $f_i \notin H$. Since $\mathcal{X} \neq \mathcal{E}$, there is a nonzero element u in KH/B. Then $f_i u \in f_i \cdot (KH/B)$, but it is clear that $(f_i u) \cdot (f_i h) \notin f_i \cdot (KH/B)$. Therefore $(f_i u) \cdot (f_i h) \neq f_i u$, whence $f_i h \notin G$, contradicting the assumption.

Thus, $G \subseteq H$. It follows now from (6) that $G = \text{Ker}\,(KH/B, H)$. Since $H = F_m$ and $B = \mathcal{X}^*(KH, H)$, we have $|H/G| = r(m)$. Hence

$$r^{\times}(n) = |F_n/G| = |H/G| \cdot |F_n/H| = r(m)j = r((n-1)f(n) + 1)f(n),$$

as required. $\square$

In a paper of Ol'shansky [72], the following two assertions on varieties of abstract groups were proved: (i) there exist distinct locally finite varieties of groups whose order functions are equal; (ii) there exists a continuum of locally finite varieties of groups whose order functions are pairwise distinct. Since for distinct varieties of groups Θ_1 and Θ_2 the varieties of representations $\omega\Theta_1$ and $\omega\Theta_2$ are also distinct, these assertions and Proposition 2.3.2 yield:

2.3.3. Corollary. (i) *Over an arbitrary field, there exist distinct locally finite varieties of group representations whose order functions are equal.*

(ii) *Over an arbitrary field, the cardinality of the set of order functions of locally finite varieties of group representations is equal to continuum.* $\square$

Let us investigate now the behavior of the order functions under the multiplication of varieties of group representations. By 2.1.3 (i), if $\mathcal{X}$ and $\mathcal{Y}$ are locally finite-dimensional varieties, then their product $\mathcal{X}\mathcal{Y}$ is also locally finite-dimensional. The following proposition gives substantially more precise information.

2.3.4. Proposition. *Let $\mathcal{X}$ and $\mathcal{Y}$ be locally finite-dimensional varieties of group representations, and let l_1, l_2 and l be the left order functions of $\mathcal{X}$, $\mathcal{Y}$ and $\mathcal{X}\mathcal{Y}$ respectively. Then*

$$l(n) = (n-1)l_1(n)l_2(n) + l_1(n) + l_2(n).$$

P r o o f. We use two well known results of the theory of firs (free ideal rings). Let $I = \mathcal{Y}^*(KF_n, F_n)$, then

$$\dim(KF_n/I) = l_2(n). \tag{7}$$

According to a theorem of Cohn [11, 12], the group algebra KF_n is a fir, i.e. all its ideals are free KF_n-modules. Moreover, it is known that an ideal of codimension d in KF_n is a free KF_n-module of rank $(n-1)d+1$ (Lewin [55]). Therefore (7) implies that I is a free KF_n-module of rank $r = (n-1)l_2(n) + 1$.

Let $e_1, \ldots, e_r$ be a basis of I over KF_n:

$$I = e_1 KF_n \oplus \cdots \oplus e_r KF_n.$$

Denote $J = \mathcal{X}^*(I, F_n)$, then

$$J = \bigoplus_{i=1}^{r} e_i \mathcal{X}^*(KF_n, F_n), \quad I/J = \bigoplus_{i=1}^{r} e_i(KF_n/\mathcal{X}^*(KF_n, F_n)),$$

whence

$$\dim(I/J) = r\, l_1(n) = ((n-1)l_2(n) + 1)\, l_1(n). \tag{8}$$

By (7) and (8),

$$\dim(KF_n/J) = (n-1)l_1(n)l_2(n) + l_2(n).$$

On the other hand, $J = \mathcal{X}^*(\mathcal{Y}^*(KF_n, F_n)) = (\mathcal{X}\mathcal{Y})^*(KF_n, F_n)$, that is, $(KF_n/J, F_n) = \mathrm{Fr}_n(\mathcal{X}\mathcal{Y})$. Therefore $\dim(KF_n/J) = l(n)$. $\square$

2.3.5. Corollary. *If $\mathcal{X}$ and $\mathcal{Y}$ are locally finite-dimensional varieties of group representations, then the left order functions of the varieties $\mathcal{X}\mathcal{Y}$ and $\mathcal{Y}\mathcal{X}$ coincide.* $\square$

In particular, for left order functions one can now give another proof of Corollary 2.3.3 (i), not depending on the theory of varieties of groups. Indeed, it is enough to take arbitrary noncommuting locally finite-dimensional varieties $\mathcal{X}$ and $\mathcal{Y}$ and consider the varieties $\mathcal{X}\mathcal{Y}$ and $\mathcal{Y}\mathcal{X}$.

The question on the right order function of the product of varieties is considerably more complicated. It makes sense only over a field of prime characteristic, for it is easy to see that over a field of characteristic 0 the product of locally finite varieties is *never* locally finite. So let $\mathcal{X}$ and $\mathcal{Y}$ be locally finite varieties of group representations over a field K of prime characteristic p. According to 2.1.3 and 2.2.9, their product $\mathcal{X}\mathcal{Y}$ is also locally finite. Denote the right order functions of $\mathcal{X}$, $\mathcal{Y}$ and $\mathcal{X}\mathcal{Y}$ by r_1, r_2 and r respectively. It is natural to ask, how does r depend on r_1 and r_2? Unlike the case of left order functions, we do not know an explicit formula expressing r in terms of r_1 and r_2; moreover, it is not known whether r_1 and r_2 determine r uniquely. However, we can claim that $r(n)$ is *bounded* by a number depending only on $r_1(n), r_2(n)$ and, of course, the characteristic of the ground field. In proving this fact, the following statement plays the main role.

2.3.6. Proposition. *Let* $\rho = (E_n, F_n)$ *be a relatively free cyclic representation of rank n over a field K of characteristic p. If E_n is finite-dimensional, then the group $F_n/\operatorname{Ker}\rho$ is finite and its order is bounded by a number depending only on n, p and $\dim E_n$.*

P r o o f. a) We may assume that K is algebraically closed. Indeed, if $\bar{K}$ is the algebraic closure of K, then it is not hard to see that $\rho_{\bar{K}} = (\bar{K} \otimes_K E_n, F_n)$ is a relatively free cyclic representation of rank n over $\bar{K}$. Also, it is evident that

$$\dim_K E_n = \dim_{\bar{K}}(\bar{K} \otimes_K E_n), \quad \operatorname{Ker}\rho = \operatorname{Ker}\rho_{\bar{K}},$$

whence the desired fact follows. Thus, in the sequel K is algebraically closed.

b) Denote $F_n/\operatorname{Ker}\rho = G, \dim E_n = d$. The group G can be considered as a matrix group of degree d. By Lemma 2.2.3, there exists a polynomial φ of degree $\leq d$ such that the eigenvalues of all matrices from G are roots of φ. Since the number of these roots does not exceed d, the set $\operatorname{tr} G = \{\operatorname{tr} g \mid g \in G\}$ is finite. It is easy to see that

$$|\operatorname{tr} G| \leq \binom{2d-1}{d}. \tag{9}$$

c) Let

$$0 = A_0 \subset A_1 \subset \cdots \subset A_k = E_n \tag{10}$$

be a G-composition series in E_n. Since dim $E_n = d$, we have $k \leq d$. Denote by H_i the kernel of the action of G on the quotient A_i/A_{i-1}. Then G/H_i is an irreducible matrix group of degree $\leq d$ over K. Since K is algebraically closed, this group is absolutely irreducible. Using Burnside's theorem (see e.g. [87, p.254]) and taking into account (9), we have

$$|G/H_i| \leq \left(\frac{2d-1}{d}\right)^{d^2}.$$

Let $H = \bigcap H_i$, then G/H is embeddable in $G/H_1 \times \cdots \times G/H_k$, whence

$$j = |G/H| \leq \left(\frac{2d-1}{d}\right)^{d^3}. \tag{11}$$

d) Since G is a group with n generators (recall that $G = F_n/\text{Ker}\,\rho$), it follows from (11) and the Schreier formula that H can be generated by a finite number of elements m satisfying the condition

$$m \leq (n-1)j + 1 \leq (n-1)\left(\frac{2d-1}{d}\right)^{d^3} + 1. \tag{12}$$

Further, H acts faithfully on E_n and stabilizes the series (10). Since it has m generators and char $K = p$, it is a finite p-group whose order is bounded by a number depending only on m, p and d (it is not hard to find this bound explicitly). By (11) and (12), the order of G is bounded by a number depending only on n, p and d. $\square$

Note. It was proved in the previous section (Corollary 2.2.9) that a locally finite-dimensional variety of group representations over a field of prime characteristic is automatically locally finite. Proposition 2.3.6 is actually a refinement of this result.

2.3.7. Corollary. *For every prime p there exists a function $f_p : \mathbb{N}^2 \to \mathbb{N}$ such that if $\mathcal{X}$ is an arbitrary locally finite variety of group representations over an arbitrary field of characteristic p, then*

$$\forall n \in \mathbb{N} : r(n) \leq f_p(n, l(n))$$

where, as usual, l and r are respectively the left and the right order functions of the variety $\mathcal{X}$. □

We can now prove the promised result concerning the right order function of the product of varieties. Let $\mathcal{X}$ and $\mathcal{Y}$ be two locally finite varieties. Denote the order functions of $\mathcal{X}$ by l_1 and r_1, the order functions of $\mathcal{Y}$ by l_2 and r_2, and the order functions of $\mathcal{X}\mathcal{Y}$ by l and r.

2.3.8. Corollary. *For every prime p there exists a function g_p : $\mathbb{N}^3 \to \mathbb{N}$ such that if $\mathcal{X}$ and $\mathcal{Y}$ are arbitrary locally finite varieties of group representations over an arbitrary field of characteristic p, then*

$$\forall n \in \mathbb{N} : \ r(n) \leq g_p(n, r_1(n), r_2(n)).$$

P r o o f. Evidently $l_1(n) \leq r_1(n)$ and $l_2(n) \leq r_2(n)$. By 2.3.4, the numbers $r_1(n)$ and $r_2(n)$ give a bound for $l(n)$; by 2.3.7, they give a bound for $r(n)$ as well. □

It was already noted that it is still unknown whether there exists an *explicit* formula expressing r in terms of r_1 and r_2. In this connection, one can naturally ask the following concrete question.

2.3.9. Problem. *Let $\mathcal{X}, \mathcal{Y}, \mathcal{X}', \mathcal{Y}'$ be locally finite varieties over a field of prime characteristic, such that the right order functions of $\mathcal{X}$ and $\mathcal{Y}$ coincide with those of $\mathcal{X}'$ and $\mathcal{Y}'$ respectively. Does the right order function of $\mathcal{X}\mathcal{Y}$ coincide with that of $\mathcal{X}'\mathcal{Y}'$?*

In investigating order functions, it is interesting and natural to estimate their growth rate. For left order functions, this question is not hard to solve. Namely, a function f of a natural argument n is said to have *polynomial growth of degree k* if $f(n) = O(n^k)$ for some natural k. Similarly, f is a function of *exponential growth* if $f(n) = O(a^n)$ for some $a > 0$.

2.3.10. Proposition. *Let $\mathcal{X}$ be a locally finite-dimensional variety. If $\mathcal{X}$ is stable of class s, then its left order function l has polynomial growth* *of degree $\leq s - 1$, but if X is not stable, then the growth of l is at least exponential.*

First we prove one auxiliary fact (cf. [68, 24.51]).

2.3.11. Lemma. *Let $\rho = (E_n, F_n)$ be a relatively free cyclic representation of rank n. If ρ is not s-stable, then*

$$\dim E_n \geq \sum_{r=1}^{m} \binom{n}{r}, \quad \text{where} \quad m = \min(s, n).$$

P r o o f. By definition, $E_n = KF_n/I_n$ with I_n a verbal ideal of KF_n. Let $x_1, \ldots, x_n$ be free generators of F_n. For every r-tuple of natural numbers $j_r = (i(1), i(2), \ldots, i(r))$, where $1 \leq r \leq m$ and $1 \leq i(1) \leq i(2) \leq \cdots \leq i(r)$, choose in KF_n the monomial

$$m(j_r) = z_{i(1)} z_{i(2)} \cdots z_{i(r)}$$

where, as usual, $z_i = x_i - 1$. We prove that the set of all monomials of this form is linearly independent modulo I_n. To do this, we introduce a linear order on the set of monomials, setting $m(j_r) < m(j_s)$ if $r < s$ and letting monomials of the same degree be ordered in an arbitrary but fixed way. Suppose there exists a relation of linear dependence

$$\sum_{j_p} \alpha_{j_p} m(j_p) \equiv 0 \tag{13}$$

where $\alpha_{j_p} \in K$ and $\equiv$ denotes equality modulo I_n. We may assume that all α_{j_p} are nonzero. Let $m(j_r)$ be the minimal monomial in (13), then (13) can be rewritten in the form

$$m(j_r) \equiv -\frac{1}{\alpha_{j_p}} \sum_{j_p \neq j_r} \alpha_{j_p} m(j_p). \tag{14}$$

In (14) let us substitute 1 for all x_i's not involved in $m(j_r)$. Then all the $m(j_p)$ from the right side will vanish, whence $m(j_r) \equiv 0$, that is, $m(j_r) = z_{i(1)} z_{i(2)} \cdots z_{i(r)} \in I_n$. Therefore the representation ρ is stable of class $r \leq s$, which is impossible.

Thus $\dim E_n$ is not less than the number of monomials of the above form. But this number, evidently, equals $\sum_{r=1}^{m} \binom{n}{r}$. $\square$

Proof of Proposition 2.3.10. If $\mathcal{X}$ is not stable, then it follows from 2.3.11 that $l(n) \geq 2^n$ for each n. Let $\mathcal{X}$ be stable of class s and let $I_n = \mathrm{Id}_n \mathcal{X}$. Then $I_n \supseteq \Delta^s$, Δ being the augmentation ideal of KF_n. Consider the series

$$KF_n \supset \Delta \supset \Delta^2 \supset \cdots \supset \Delta^s.$$

Since elements of the form $z_{i_1} \ldots z_{i_k} + \Delta^{k+1}$ form a K-basis of Δ^k/Δ^{k+1}, we have

$$\dim(KF_n/\Delta^s) = \dim(KF_n/\Delta) + \dim(\Delta/\Delta^2) + \cdots + \dim(\Delta^{s-1}/\Delta^s)$$
$$= 1 + n + \cdots + n^{s-1},$$

whence $l(n) = \dim(KF_n/I_n) \leq 1 + n + \cdots + n^{s-1}$. Thus the function l has polynomial growth of degree $\leq s - 1$. $\square$

Of course, there exist varieties of group representations whose order functions are of *hyperexponential* growth: one can mention, for instance, any variety of the form $\omega\Theta$ where Θ is a group variety of hyperexponential growth (the existence of such varieties of groups follows from results of Peter M. Neumann [68, 24.51-53]).

We note in conclusion that one can also estimate the growth of right order functions, but the corresponding results will not have such a finished form as Proposition 2.3.10. Indeed, by Theorem 2.2.2, every locally finite variety $\mathcal{X}$ of representations is contained in some $\omega\Theta$, where Θ is a locally finite variety of groups. Hence we have inequalities

$$l(n) \leq r(n) \leq f(n)$$

(which become equalities for $\mathcal{X} = \omega\Theta$ — see 2.3.1). If the growth of f is known, it is possible, at least in principle, to estimate the growth of r using Propositions 2.3.6 and 2.3.10.

2.4. Critical representations

In 1960, D.C Cross introduced the notion of a critical group. Very soon this notion was carried over to other algebraic structures and began to play a crucial role in the study of locally finite varieties (see e.g. [69, 44–46, 50, 56]). The theory of varieties of group representations is no exception, and the proofs of many important results in the field are based on the technique of critical representations.

In the present section, following [92, 93], we establish a number of basic properties of critical group representations. Most of these properties are similar to those of critical groups, rings, etc., and therefore several routine proofs are omitted. It should be noted that although some results of this section are of independent interest, their applications in the next chapter of the book will be especially important.

2.4.1. Definition. *A representation of a group is called* critical *if it is finite and is not contained in the variety generated by its proper factors.*

Here, as usual, by a *factor* (or *section*) of a representation ρ we mean an epimorphic image of some subrepresentation σ of ρ; this factor is *proper* unless $\sigma = \rho$ and the epimorphism is identical.

2.4.2. Lemma. *Every locally finite variety is generated by its critical representations. Every finite representation belongs to the variety generated by its critical factors.*

P r o o f. We prove the first assertion of the lemma. If $\mathcal{X}$ is a locally finite variety, it is generated by its finite representations. Let $\mathcal{Y}$ be the variety generated by all critical representations from $\mathcal{X}$. Suppose that $\mathcal{X} \neq \mathcal{Y}$, then there exists a finite representation in $\mathcal{X}$ not belonging to $\mathcal{Y}$. Let ρ be such a representation of the smallest possible order. The order of each proper factor of ρ is strictly less than that of ρ, hence this factor is contained in $\mathcal{Y}$. Therefore ρ is a critical representation, which is impossible in view of the definition of $\mathcal{Y}$.

The proof of the second assertion is similar. □

Let $\rho = (V, G)$ be an arbitrary representation. The *socle* $\mu(\rho) = \mu(V, G)$ of ρ is the sum of all minimal G-submodules of V. If there are no minimal G-submodules in V, we set $\mu(\rho) = \{0\}$. On the other hand, denote by $\kappa(\rho) = \kappa(V, G)$ the *radical* of ρ, that is, the intersection of all maximal G-submodules of V. If there are no maximal G-submodules in V, we set $\kappa(\rho) = V$.

A representation $\rho = (V, G)$ is called *monolithic* if V has a unique minimal nonzero G-submodule, called the *monolith* of ρ. Dually, ρ is *comonolithic* if V has a unique maximal proper G-submodule which is called the *comonolith* of ρ. It is obvious that if ρ has a monolith or a comonolith, then it coincides with $\mu(\rho)$ or $\kappa(\rho)$ respectively.

Denote by $(QS\text{-}1)\rho, (S\text{-}1)\rho$ and $(Q\text{-}1)\rho$ the sets of all proper factors, all proper subrepresentations and all proper epimorphic images of ρ respectively. Let ρ be a finite representation. According to the above definition, ρ is critical if $\rho \notin \mathrm{var}((QS\text{-}1)\rho)$. Similarly, ρ is said to be S-*critical* if $\rho \notin \mathrm{var}((S\text{-}1)\rho)$ and Q-*critical* if $\rho \notin \mathrm{var}((Q\text{-}1)\rho)$.

2.4.3 Lemma. *Every Q-critical representation is monolithic, and every S-critical representation is comonolithic. In particular, a critical representation is both monolithic and comonolithic.*

P r o o f. The first assertion is routine. As to the second, consider an arbitrary S-critical representation (V, G) and let $W = \kappa(V, G)$. Then V/W is a completely reducible G-module. Consider its decomposition $V/W = \bigoplus_{i=1}^{n}(V_i/W)$ into a sum of irreducible G-submodules. Since $V = \sum_{i=1}^{n} V_i$, the representation (V, G) is contained in the radical class generated by its subrepresentations (V_i, G) and therefore in the variety generated by the (V_i, G). If $n > 1$, all these subrepresentations are proper, and so (V, G) cannot be S-critical. Therefore $n = 1$, whence V/W is irreducible and thus W is the comonolith of (V, G). $\square$

2.4.4. Note. It should be noted that the first part of Lemma 2.4.3 can be reversed, i.e. a finite faithful representation is Q-critical if and only if it is monolithic [92]. This is completely analogous to a result from [45]. However, for the second part of the lemma the converse is not true: *there exists a* *comonolithic representation which is not S-critical.* Indeed, let $K = \mathbb{F}_{p^n}$ where p is a prime and $n \geq 2$. Consider the representation $\rho = (V, G)$ where V is a two-dimensional K-space with basis e_1, e_2 and

$$G = \mathrm{UT}_2(K) = \begin{pmatrix} 1 & 0 \\ K & 1 \end{pmatrix}$$

acts on V in a natural way. Then ρ is comonolithic since $\langle e_1 \rangle$ is the only nontrivial G-submodule of V. Take in G the subgroup

$$H = \begin{pmatrix} 1 & 0 \\ \mathbb{F}_p & 1 \end{pmatrix},$$

then (V, H) is a proper subrepresentation of ρ. Clearly $\mathcal{S} \subset \mathrm{var}\,(V, H) \subseteq \mathcal{S}^2$. Since there are no intermediate varieties between $\mathcal{S}$ and $\mathcal{S}^2$ (an easy exercise!), we have $\mathrm{var}\,(V, H) = \mathcal{S}^2$ and, consequently, $\rho \in \mathrm{var}\,(V, H)$. Hence ρ is not S-critical.

Our subsequent considerations are based on the so-called minimal realization of a finite object in a locally finite variety going back to Kovács and Newman [45]. First of all, we state one simple fact which follows directly from Lemma 2.1.5.

2.4.5. Lemma. *Let $\mathcal{D}$ be a class of representations. If the variety $\mathcal{X} = \mathrm{var}\,\mathcal{D}$ is locally finite-dimensional, then all finitely generated representations from $\mathcal{X}$ belong to the class $\mathrm{VQSD}_0\,\mathcal{D}'$, where $\mathcal{D}'$ is the class of all faithful images of representations from $\mathcal{D}$.*

Proof. Since $\mathcal{D} \subseteq \mathrm{V}\mathcal{D}'$, Lemma 2.1.5 implies that $\mathcal{X} \subseteq \mathrm{VQSD}_0\mathrm{V}\mathcal{D}'$. The rest follows from the obvious relations $\mathrm{D}_0\mathrm{V} \leq \mathrm{VD}_0, \mathrm{SV} \leq \mathrm{VS}, \mathrm{QV} \leq \mathrm{VQ}$ between the closure operations. $\square$

Let now $\mathcal{D}$ be a class of finite representations closed with respect to taking factors, and let $\mathrm{var}\,\mathcal{D}$ be locally finite. Take an arbitrary finite representation $\rho = (V, G)$ in $\mathrm{var}\,\mathcal{D}$. By the previous lemma, $\rho \in \mathrm{VQSD}_0\mathcal{D}'$. Therefore if $\bar{\rho} = (V, \bar{G})$ is the faithful image of ρ, then $\bar{\rho} \in \mathrm{QSD}_0\mathcal{D}$. Thus the faithful image of any finite representation from $\mathrm{var}\,\mathcal{D}$ can be presented as a factor of the direct product of a finite number of faithful representations from $\mathcal{D}$. We will call this a *realization of* $\rho = (V, G)$ *in* var $\mathcal{D}$.

From all realizations of $\rho = (V, G)$ in var $\mathcal{D}$, we choose a minimal one in the following sense. Each realization determines a non-increasing sequence of positive integers consisting of the orders of factors occurring in the direct product. If we order these sequences lexicographically on the left, then the set of all sequences will have the first element. The corresponding realization is called a *minimal realization of the given* $\rho = (V, G)$ *in* var $\mathcal{D}$. Of course, it is not uniquely determined. Let us fix one of the minimal realizations of $\rho = (V, G)$ in var $\mathcal{D}$:

$$(V, \bar{G}) = (B/A,\ S/R), \text{ where } (A, R) \lhd (B, S) \subseteq \prod (D_i, H_i), \qquad (1)$$

each (D_i, H_i) being a representation from $\mathcal{D}$. The following three lemmas deal with this fixed realization. Their proofs are standard and straightforward (cf. [45] or [68, 53.21-26]) and therefore are omitted.

2.4.6. Lemma. *Every representation* (D_i, H_i) *is critical.* $\square$

Denote by π_i the natural projection of the direct product $\prod (D_i, H_i)$ onto its i-th factor (D_i, H_i).

2.4.7. Lemma. *The representation* (B, S) *is a subdirect product in* $\prod (D_i,\ H_i)$, *that is,* $(B, S)^{\pi_i} = (D_i, H_i)$ *for every* i. *The representation* (A, R) *has the trivial intersection with every* (D_i, H_i). *If* W *is a nonzero* H_i-*submodule of* D_i, *then* $W \cap B \neq 0$. $\square$

Since each (D_i, H_i) is critical, by Lemma 2.4.3 it is monolithic. Let $M_i = \mu(D_i, H_i)$ be its monolith. With a slight abuse of language, sometimes the corresponding representation (M_i, H_i) will be also called the monolith of (D_i, H_i). Recall that representations ρ and σ are said to be equivalent (notation $\rho \sim \sigma$) if their faithful images are isomorphic.

2.4.8. Lemma. *For every* i *there exists a minimal* G-*submodule* N_i *of* V *such that* $(N_i, G) \sim (M_i, H_i)$. *In particular, if* (V, G) *is monolithic, then all* (M_i, H_i) *are equivalent to the monolith* (M, G) *of the given* (V, G). *Furthermore,* $(V/M,\ G) \in \text{var}\{(D_i/M_i, H_i)\,|\ i = 1, 2, \dots\}$. $\square$

The following theorem is parallel to a well known group-theoretic result of Kovács and Newman [45, 46], but in proving it one has to overcome additional difficulties. These difficulties stem from the fact that finite representations of groups over an *infinite* field are not, strictly speaking, completely "finite" objects.

2.4.9. Theorem. *A representation is critical if and only if it is both S-critical and Q-critical.*

First we establish one auxiliary fact.

2.4.10. Lemma. *Let $\rho = (V, G)$ be a finite representation, $\mathcal{X}$ a proper subvariety of* var ρ *and* $\phi = (A, R)$ *a finite representation in* var $\rho \setminus \mathcal{X}$. *Then* var ρ *contains a subvariety* $\mathcal{Y}$ *and a representation* σ *such that*

(a) $\mathcal{X} \subseteq \mathcal{Y}$;

(b) σ *is a critical factor of* ρ;

(c) $(QS\text{-}1)\sigma \subseteq \mathcal{Y}$;

(d) $\phi \notin \mathcal{Y}$ *but* $\phi \in \mathcal{Y} \vee$ var σ.

Proof. Let $\mathcal{C}$ be a set of all critical factors of ρ. For each positive integer i denote by $\mathcal{C}_i$ the subset of all representations from $\mathcal{C}$ whose orders do not exceed i. Then

$$\mathcal{C}_1 \subseteq \mathcal{C}_2 \subseteq \cdots \subseteq \mathcal{C}_t = \mathcal{C},$$

where $t = |\rho|$. Denote $\mathcal{Y}_i = \mathcal{X} \vee$ var $\mathcal{C}_i$ for each $i = 1, \ldots, t$. Then

$$\mathcal{X} = \mathcal{Y}_1 \subseteq \mathcal{Y}_2 \subseteq \ldots \mathcal{Y}_t = \text{var } \rho.$$

Since $\phi \in$ var $\rho \setminus \mathcal{X}$, there exists i such that $\phi \notin \mathcal{Y}_i$ but $\phi \in \mathcal{Y}_{i+1}$. Let $\mathcal{X}_0$ be the class of all critical representations from $\mathcal{X}$ and let $\mathcal{D} = \mathcal{X}_0 \cup \mathcal{C}_{i+1}$. Evidently var $\mathcal{D} = \mathcal{Y}_{i+1}$. Furthermore, every proper critical factor of every representation from $\mathcal{D}$ is contained in $\mathcal{X}_0 \cup \mathcal{C}_i \subseteq \mathcal{Y}_i$ (because the order of a proper factor of a finite faithful representation is strictly less than the order of the given representation).

Denote $\bar{\mathcal{D}} = QS\mathcal{D}$, then $\bar{\mathcal{D}}$ is closed under taking factors. Since $\phi \in \mathcal{Y}_{i+1} =$ var $\bar{\mathcal{D}}$, there is a minimal realization of $\phi = (A, R)$ in var $\bar{\mathcal{D}}$:

$$(A, \bar{R}) = (W_1/W_2, T_1/T_2), \text{ where } (W_1, T_1) \subseteq \prod_{j=1}^{n} (D_j, H_j).$$

All (D_j, H_j) are critical representations belonging to $\bar{\mathcal{D}}$, therefore it follows from the above that each (D_j, H_j) is contained either in $\mathcal{X}_0 \cup \mathcal{C}_i$ or in $\mathcal{C}_{i+1} \setminus (\mathcal{X}_0 \cup \mathcal{C}_i)$. We may assume that

$$(D_j, H_j) \in \mathcal{X}_0 \cup \mathcal{C}_i \ \text{ if } \ 1 \leq j \leq k \ ,$$

$$(D_j, H_j) \in \mathcal{C}_{i+1} \setminus (\mathcal{X}_0 \cup \mathcal{C}_i) \ \text{ if } \ k+1 \leq j \leq n \ .$$

Since $\phi \notin \mathcal{Y}_i = \text{var}(\mathcal{X}_0 \cup \mathcal{C}_i)$, the set $(D_{k+1}, H_{k+1}), \ldots, (D_n, H_n)$ is nonempty. There exists m such that

$$\phi \notin \text{var}\{\mathcal{X}, \mathcal{C}_i, (D_{k+1}, H_{k+1}), \ldots, (D_m, H_m)\},$$

$$\phi \in \text{var}\{\mathcal{X}, \mathcal{C}_i, (D_{k+1}, H_{k+1}), \ldots, (D_{m+1}, H_{m+1})\}.$$

Denote $\text{var}\{\mathcal{X}, \mathcal{C}_i, (D_{k+1}, H_{k+1}), \ldots, (D_{m+1}, H_{m+1})\} = \mathcal{Y}$, (D_{m+1}, H_{m+1}) $= \sigma$. Then for $\mathcal{Y}$ and σ the conditions (a), (b) and (d) are satisfied. It remains to prove that (c) is also satisfied.

Let σ_1 be a proper factor of σ. All critical factors of σ_1 are critical factors of the initial representation ρ, and their orders are strictly less than $|\sigma| = i + 1$. Therefore all critical factors of σ_1 are contained in $\mathcal{C}_i \subseteq \mathcal{Y}$, whence $\sigma_1 \in \mathcal{Y}$. $\square$

Proof of Theorem 2.4.9. To prove the nontrivial part, suppose that $\rho = (V, G)$ is both S-critical and Q-critical (in particular, ρ is finite and faithful). If $\mathcal{X} = \text{var}(\text{S-1})\rho$, then $\rho \notin \mathcal{X}$ and, by the previous lemma, there exist a subvariety $\mathcal{Y}$ and a representation σ of var ρ such that

(a) $\mathcal{X} \subseteq \mathcal{Y}$;
(b) σ is a critical factor of ρ;
(c) $(\text{QS-1})\{\sigma\} \subseteq \mathcal{Y}$;
(d) $\rho \notin \mathcal{Y}$ but $\rho \in \text{var} \, \sigma \vee \mathcal{Y}$.

Denote by $\mathcal{D}_0$ the class of all finite representations from $\mathcal{Y}$ and let $\mathcal{D} = \mathcal{D}_0 \cup \{\sigma\}$. Then $\mathcal{D}$ is factor-closed and var ρ = var $\mathcal{D}$. Let

$$\rho = (W_1/W_2, T_1/T_2), \text{ where } (W_1, T_1) \subseteq \prod_{j=1}^{n} (D_j, H_j)$$

be a minimal realization of the (faithful!) representation ρ in var $\mathcal{D}$. All the (D_j, H_j) are critical and, since $\rho \notin \mathcal{Y}$, some of them must coincide with σ. Thus we have

$$(W_1, T_1) \subseteq (D_1, H_1) \times \cdots \times (D_s, H_s) \times \prod_{i=1}^{t} (B_i, S_i) = (W, T)$$

where $(D_j, H_j) \in \mathcal{D}_0 \subseteq \mathcal{Y}$ and each (B_i, S_i) is isomorphic to $\sigma = (B, S)$. Being critical, (B_i, S_i) has a monolith $M_i = \mu(B_i, S_i)$. By (c), $M_i = \mathcal{Y}^*(B_i, S_i)$ and so $\mathcal{Y}^*(W, T) = \bigoplus_{i=1}^{t} M_i$. Since $\mathcal{Y}^*(W_1, T_1) \subseteq \mathcal{Y}^*(W, T)$, we have

$$\mathcal{Y}^*(\rho) = \mathcal{Y}^*(W_1/W_2, T_1/T_2) \subseteq ((\bigoplus M_i) + W_2)/W_2 .$$

Each summand $(M_i + W_2)/W_2$ is invariant and irreducible with respect to $G = T_1/T_2$, therefore the representation $(\mathcal{Y}^*(\rho), G)$ is completely reducible and is decomposed into a direct sum of G-representations which are equivalent to $(\mu(\sigma), S)$.

On the other hand, ρ is monolithic, hence by Lemma 2.4.8

$$(\mu(\rho), G) \sim (\mu(\sigma), S).$$

Together with the above, this implies that $\mathcal{Y}^*(\rho) = \mu(\rho)$. Therefore $(\text{Q-1})\rho \subseteq \mathcal{Y}$. Consequently, since $(\text{S-1})\rho \subseteq \mathcal{Y}$, it follows that $(\text{QS-1})\rho \subseteq \mathcal{Y}$. By (d), ρ is critical. $\square$

2.5. Critical representations and irreducibility

As we said at the outset, the main applications of the technique of critical representations will be demonstrated in Chapter 3. Nevertheless, it is appropriate to present some applications right now. In particular, we will see that there is a close connection between criticality of a representation and its irreducibility. This connection becomes absolutely transparent if the corresponding representation is ordinary.

Recall that a representation of a finite group G is said to be *ordinary* if the order of G is not divisible by the characteristic of the ground field, and *modular* otherwise. Further, we will say that a representation is *simple* if it is faithful and irreducible simultaneously (this terminology was introduced in [78] and is justified by the following obvious observation: a representation is simple if and only if it does not have proper epimorphic images).

2.5.1. Proposition. *A finite simple representation is critical. A finite ordinary representation is critical if and only if it is simple.*

P r o o f. Let $\rho = (V, G)$ be a finite simple representation and let $\mathcal{D}$ be the set of its proper factors. Suppose that $\rho \in \mathrm{var}\, \mathcal{D}$. Let

$$(V, \bar{G}) = (B/A,\ S/R), \text{ where } (B, S) \subseteq \prod(D_i, H_i),\ (D_i, H_i) \in \mathcal{D}$$

be a minimal realization of ρ in $\mathrm{var}\, \mathcal{D}$. Since ρ coincides with its monolith, by Lemma 2.4.8 it is equivalent to the monolith of each (D_i, H_i). But the order of (D_i, H_i) is strictly less then the order of ρ, which gives a contradiction. Thus $\rho \notin \mathrm{var}\, \mathcal{D}$, that is ρ is critical.

Let now ρ be an ordinary critical representation. By the Maschke Theorem, it is completely reducible, and since ρ is monolithic, it must be irreducible. $\square$

Of course, a modular critical representation need not be irreducible — see the examples at the end of this section.

2.5.2. Proposition. *Let $\rho = (V, G)$ and $\sigma = (W, H)$ be critical representations such that $\mathrm{var}\, \rho = \mathrm{var}\, \sigma$. Then the monoliths of ρ and σ are equivalent.*

P r o o f. Denote by $\mathcal{D}$ the set of all factors of σ. Then the variety $\mathcal{X} = \mathrm{var}\, \rho = \mathrm{var}\, \sigma$ is generated by $\mathcal{X}$. Consider a minimal realization of ρ in $\mathrm{var}\, \mathcal{D}$:

$$\rho = (B/A,\ S/R), \text{ where } (B, S) \subseteq \prod(D_i, H_i) \in \mathcal{D}.$$

If all the (D_i, H_i) were proper factors of σ, the latter could not be critical. Therefore $\sigma = (D_i, H_i)$ for some i. But then Lemma 2.4.8 guarantees that the monoliths of ρ and σ are equivalent. $\square$

From Propositions 2.5.1 and 2.5.2 we immediately deduce the following elegant result:

2.5.3. Corollary. *Two finite simple representations generate the same variety if and only if they are isomorphic.* $\square$

In other words, *a faithful irreducible representation of a finite group is uniquely determined by its identities.* Although this result is not difficult, it provides a strong motivation for the study of identities of group representations, especially when finite groups are concerned.

The next question was raised in [76]: for which varieties of groups Θ the lattice of subvarieties in $\omega\Theta$ is distributive? The above technique makes it possible to find a wide class of group varieties for which the answer is positive. Let us agree to say that a class of finite simple representation $\mathcal{D}$ is *closed* if each simple factor of a representation from $\mathcal{D}$ also belongs to $\mathcal{D}$.

2.5.4. Theorem (Plotkin [78], Vovsi [92]). *Let Θ be a locally finite variety of groups whose exponent is not divided by the characteristic of the ground field. Then all subvarieties in $\omega\Theta$ are in one-to-one correspondence with closed classes of finite simple representations from $\omega\Theta$.*

P r o o f. For a subvariety $\mathcal{X}$ of $\omega\Theta$, let $\mathcal{X}'$ be the class of all finite simple representations from $\mathcal{X}$. Evidently $\mathcal{X}'$ is closed. On the other hand, if $\mathcal{D}$ is a closed class of finite simple representations from $\omega\Theta$, we set $\mathcal{D}' = \operatorname{var}\mathcal{D}$. Let us show that the maps $\mathcal{X} \mapsto \mathcal{X}'$ and $\mathcal{D} \mapsto \mathcal{D}'$ are mutually inverse bijections, that is, $\mathcal{X}'' = \mathcal{X}$ and $\mathcal{D}'' = \mathcal{D}$.

Let $\mathcal{X} \subseteq \omega\Theta$. Since $\mathcal{X}$ is locally finite, it is generated by finite representations. These representations are ordinary and therefore comletely reducible. It follows that $\mathcal{X}$ is generated by its finite simple representations, whence $\mathcal{X}'' = \mathcal{X}$.

Let now $\mathcal{D}$ be a closed class of finite simple representations from $\omega\Theta$. Denote $\mathcal{D}_0 = \mathrm{QS}\mathcal{D}$, then $\mathcal{D}_0$ is closed under taking factors, each simple representation from $\mathcal{D}_0$ is contained in $\mathcal{D}$, and $\operatorname{var}\mathcal{D}_0 = \operatorname{var}\mathcal{D}$. If ρ is a finite simple representation from $\operatorname{var}\mathcal{D}_0$, we choose a minimal realization of ρ in $\operatorname{var}\mathcal{D}_0$:

$$\rho = (B/A,\ S/R), \quad \text{where} \quad (B,S) \subseteq \prod(D_i, H_i),\ D_i, H_i \in \mathcal{D}_0.$$

By 2.4.8, the monoliths of ρ and (D_i, H_i) are equivalent. But ρ coincides with its monolith, while the monolith of (D_i, H_i) belongs to $\mathcal{D}$ since $(D_i, H_i) \in \mathcal{D}_0$. Consequently, $\rho \in \mathcal{D}$. Thus, all finite simple representations from $\mathrm{var}\,\mathcal{D} = \mathcal{D}'$ belong to $\mathcal{D}$, whence $\mathcal{D}'' = \mathcal{D}$. $\square$

2.5.5. Corollary. *If Θ is a locally finite variety of groups whose exponent is not divisible by the characteristic of the ground field, then the lattice of subvarieties in $\omega\Theta$ is distributive.*

Indeed, it follows from the previous theorem that this lattice is isomorphic to the lattice of all closed classes of finite simple representations from the variety $\omega\Theta$. $\square$

Notes. 1. It should be emphasized that the lattice of subvarieties of a locally finite variety of group Θ need not be distributive. At the same time, we have proved that if $\mathrm{char}\,K \nmid \exp\Theta$, then the lattice of subvarieties of $\omega\Theta$ is *always* distributive. This is rather curious, because there is a natural injection (but not a monomorphism!) of the lattice of subvarieties of Θ into the lattice of subvarieties of $\omega\Theta$, so that the latter is, in a sense, "larger".

2. Under the same conditions on Θ, the lattice of subvarieties of $\omega\Theta$ is finite if and only if Θ is abelian. This fact will be proved in § 3.5.

The following result contains Proposition 2.5.2 and shows that if two critical representations have the same identities, then a number of their structural properties are identical.

2.5.6. Theorem (Vovsi [93]). *Let $\rho = (V, G)$ and $\sigma = (W, H)$ be critical representations such that $\mathrm{var}\,\rho = \mathrm{var}\,\sigma$. Then:*
  (a) $(V/\kappa(\rho), G) \sim (W/\kappa(\sigma), H)$;
  (b) $G/\Phi(G) \simeq H/\Phi(H)$;
  (c) $(\mu(\rho), G) \sim (\mu(\sigma), H)$;
  (d) $\mathrm{var}((\text{S-1})\rho) = \mathrm{var}((\text{S-1})\sigma)$;
  (e) $\mathrm{var}((\text{Q-1})\rho) = \mathrm{var}((\text{Q-1})\sigma)$;
  (f) $\mathrm{var}((\text{QS-1})\rho) = \mathrm{var}((\text{QS-1})\sigma)$.

Here $\Phi(G)$ is the Frattini subgroup of G while all the other notation was introduced earlier. We note that Theorem 2.5.8 is similar to a well known result of Bryant [8] on critical groups. The proof in essence follows that of [8] with only a few additional considerations.

It is not an exaggeration to say that there are few results on locally finitevarieties whose proofs do not depend on critical objects. Therefore the study of critical representations will lead to a deeper understanding of a field as a whole, an understanding which will very likely open up new research directions. However, although the structure of ordinary critical representations is quite transparent in view of Proposition 2.5.1, in the modular case our knowledge is rather poor. The situation would be much better if we had a sufficient number of concrete examples of modular critical (reducible) representations. Unfortunately, at the present moment only a few isolated examples are known. Therefore an important general problem in the field is to construct new series of modular critical representations.

One approach to this problem is based on a classical notion. Let G be a finite group and let

$$KG = M_1 \oplus M_2 \oplus \cdots \oplus M_r$$

be the decomposition of the algebra KG into the direct product of indecomposable G-modules. These modules, the *principal indecomposables* of G, play a fundamental role in the modular representation theory. It is well known that they are both monolithic and comonolithic. Together with 2.4.3, this immediately suggests the following question.

2.5.7. Problem. *Are the representations of a finite group G corresponding to its principal indecomposables critical? In particular, is it true if $G = S_n$? If the answer in general is negative, find necessary and/or sufficient conditions under which it is the case.*

Another approach to constructing critical representations is based on the technique of triangular products (see [80] or [94] for the definition). The point is that if we take two irreducible representations ρ and σ, then their triangular product $\rho \bigtriangledown \sigma$ is both monolithic and comonolithic, which is already rather close to criticality. Without going into details, we note that the first two of the following examples are of this type.

Examples. 1. Let $K = \mathbb{F}_p$ and let

$$\mathrm{ut}_2(K) = (K^2, \mathrm{UT}_2(K))$$

be the canonical unitriangular representation of degree 2. Since $\mathrm{UT}_2(K)$ is a cyclic group of order p, it is easy to see that each proper factor of $\mathrm{ut}_2(K)$ is trivial, i.e. belongs to S. Evidently $\mathrm{ut}_2(K) \notin S$, and hence this representation is critical. In particular, this shows that a modular critical representation need not be irreducible.

2. Let $K = \mathbb{F}_3$ and let $\rho = (V, G)$ be the natural 2-dimensional representation of the group G of all matrices

$$\begin{pmatrix} b & 0 \\ a & 1 \end{pmatrix}, \quad \text{where} \quad a \in K,\ b \in K \setminus \{0\}.$$

It is easy to show that

$$[x_1, x_2] - 1 \tag{1}$$

is an identity of every proper factor of ρ. Indeed, since V has only one proper G-submodule, which we denote by A, it suffices to prove this assertion for factors of the following three types:

(i) (V, H) where H is a proper subgroup of G;
(ii) (A, G);
(iii) $(V/A, G)$.

The first is evident because G is a group of order 6 and each its proper subgroup is abelian. As to the representations (ii) and (iii), they both are one-dimensional and so G act there as an abelian group. Hence both these representations satisfy (1).

On the other hand, the group G is not abelian and therefore (1) is not an identity of ρ. Thus ρ is critical.

3. Let $\mathrm{Per}\, S_3 = (K^3, S_3)$ be the permutational representation of the symmetric group S_3 (i.e. S_3 acts on the 3-dimensional space $K^3 = V$ by permuting its basis vectors e_1, e_2, e_3), and suppose that $\mathrm{char}\, K$ is "bad",

that is, equals 2 or 3. Is this representation critical? Take in V two S_3-submodules

$$A = \{\alpha(e_1 + e_2 + e_3) \,|\, \alpha \in K\}, \quad B = \{\alpha e_1 + \beta e_2 + \gamma e_3 \,|\, \alpha + \beta + \gamma = 0\}$$

and consider separately two cases.

a) Let char $K = 2$. In this case $A \cap B = \emptyset$. Indeed, if $a \in A \cap B$, then $a = \alpha(e_1 + e_2 + e_3)$ where $3\alpha = 0$, and so $\alpha = 0$ and $a = 0$. Since $\dim A = 1$ and $\dim B = 2$, it follows that $V = A \oplus B$. Thus V is a decomposable S_3-module and so Per S_3 is not critical.

b) Now let char $K = 3$. We will show that in this case Per S_3 is a critical representation. More exactly, we will show that

$$([x_1, x_2] - 1)^2 \qquad (2)$$

is not an identity of Per S_3, but is an identity of every proper factor of this representation. The first is easy: choose in S_3 the transpositions $\sigma = (12)$ and $\tau = (13)$, and verify that $e_1 \circ ([\sigma, \tau] - 1)^2 = e_1 + e_2 + e_3 \neq 0$. To prove the second assertion, we note that if char $K = 3$, then $A \subset B$, and that A and B are the only proper S_3-submodules of V. Since all factors of the composition series $0 \subset A \subset B \subset V$ are one-dimensional, S_3 acts on these factors as an abelian group, and therefore the corresponding representations satisfy the identity $[x_1, x_2] - 1$. To show that (2) is an identity of every proper factor of Per S_3, it is enough to consider factors of the following types:

(i) (V, H) where H is a proper subgroup of S_3;

(ii) (B, S_3);

(iii) $(V/A, S_3)$.

But for these factors the claim is evident because: (i) every proper subgroup of S_3 is abelian; (ii)-(iii) the representations (A, S_3), $(B/A, S_3)$ and $(V/B, S_3)$ satisfy the identity $[x_1, x_2] - 1$.

2.5.8. Problem. Let Per $S_n = (K^n, S_n)$ be the permutational representation of the group S_n. For which K and n is this representation critical?

Chapter 3

IDENTITIES OF FINITE AND STABLE-BY-FINITE REPRESENTATIONS

The problem of determining which varieties are finitely based is one of the major problems of the theory of varieties of arbitrary algebraic structures. The theory of varieties of group representations is no exception, and at all stages of its development the problems associated with the existence of a finite basis of identities have remained at the center of attention. It should be emphasized, however, that the *existence* of non-finitely-based varieties of group representations is an immediate consequence of the existence of non-finitely-based varieties of abstract groups, established in 1970 by Ol'shansky [70], Adjan [1] and Vaughan-Lee [88]. Therefore the essence of the finite basis problem for varieties of group representations is the search for various interesting and natural cases in which the problem has a positive solution.

The considerations of the present chapter were inspired by the remarkable theorem of Oates and Powell [69] asserting that every finite group has a finite basis of identities. Our initial question is the following: *does every representation of a finite group over a field have a finite basis of identities?* The first step in this direction has been done in [91, 92]: it was proved that for *ordinary* representations the answer is positive. This rather simple fact was substantially generalized by Plotkin [78] who proved that there is a finite basis of identities for every *special* representation, that is, a representation $\rho = (V, G)$ such that (i) V is finite-dimensional, (ii) G has a normal subgroup of finite index acting stably on V, and (iii) $|G/H|$ is not divisible by the characteristic of the ground field.

At the same time, the question of the existence of a finite basis for identities of an *arbitrary* representation of a finite group has remained open for a rather long time. Moreover, the results of [92] and [78] motivated the following more general problem: *does every stable-by-finite representation (i.e. a representation satisfying only (ii)) have a finite basis of identities?* In 1987, this problem was solved in the affirmative by Nguyen Hung Shon and the author [101]. The main aim of the present chapter is to provide a complete proof of this result.

It should be noted that the finite basis problem for stable-by-finite representations is parallel to the well known problems on the existence of finite bases for identities of nilpotent-by-finite groups and rings. For associative rings the problem was solved quite a long time ago, but for groups and Lie rings they have only been solved with some additional restrictions [14, 90]; in general they still remain open.

The chapter is organized as follows. In § 3.1 we prove that an ordinary representation of a finite group has a finite basis of identities. Although this fact is contained in the more general results of subsequent sections, we decided to provide its proof separately because it is quite transparent and illustrative. Also, in this specific case we are able to prove a rather stronger result. Sections 3.2–3.4 are entirely devoted to proving the finite basis property for stable-by-finite representations (Theorem 3.4.2). Several corollaries and related facts are established in § 3.5.

Throughout the chapter, our ground ring K is a field.

3.1. Ordinary representations of finite groups

The ideology of the proof presented in this section goes back to Oates–Powell [69] and Kovács–Newman [44] and is based on two principal concepts: critical representation and Cross variety. By analogy with varieties of groups, we say that a variety $\mathcal{X}$ of group representations is *Cross* if (i) $\mathcal{X}$ is locally finite, (ii) $\mathcal{X}$ is finitely based, and (iii) $\mathcal{X}$ contains only a finite number of nonisomorphic critical representations. The following assertion is standard.

3.1.1. Lemma. *A subvariety of a Cross variety is also a Cross variety.*

P r o o f. Let $\mathcal{Y}$ be a subvariety of a Cross variety $\mathcal{X}$. Then $\mathcal{Y}$ certainly satisfies (i) and (iii). Further, denote $I = \mathrm{Id}\,\mathcal{X}$ and $J = \mathrm{Id}\,\mathcal{Y}$. Since $\mathcal{X}$ is finitely based, I is finitely generated as a completely invariant ideal. By (ii), $\mathcal{X}$ has a finite number of subvarieties whence, in particular, all strictly ascending chains of completely invariant overideals of I are finite. A usual "noetherian-type" argument now shows that all these overideals are finitely generated as completely invariant ideals. In particular, J is finitely generated, that is, $\mathcal{Y}$ is finitely based. $\square$

3.1.2. Theorem (Vovsi [92]). *The variety generated by an ordinary representation of a finite group is a Cross variety.*

To prove this assertion, we need two lemmas.

3.1.3. Lemma. *Let $\rho = (V, G)$ be a simple representation. If $0 \neq v \in V$ and $1 \neq g \in G$, then there exists $h \in G$ such that $v \circ (1 - g^h) \neq 0$.*

P r o o f. Suppose that $v \circ (1 - g^h) = v - v \circ (h^{-1}gh) = 0$ for every $h \in G$. Then for every $h \in G$

$$v \circ h^{-1} - (v \circ h^{-1}) \circ g = 0. \qquad (1)$$

Let B be the G-submodule of V generated by all $v \circ h^{-1}$, $h \in G$. Since $v \neq 0$, we have $B \neq \{0\}$, and since ρ is irreducible, it follows that $B = V$. Now (1) implies that $g \in \mathrm{Ker}\,\rho$ which is impossible because ρ is faithful. $\square$

For group representations, we define now an analogue of the so-called chief centralizer identity from [69]. We set

$$w_1 = 1 - (x_0^{-1}x_1)^{x_{01}}$$

and inductively

$$w_d = w_{d-1}(1 - (x_0^{-1}x_d)^{x_{0d}})\ldots(1 - (x_{d-1}^{-1}x_d)^{x_{d-1,d}})$$

where x_i, x_{ij} are pairwise distinct free generators of F. Then w_d belongs to KF and involves

$$(d+1) + 1 + 2 + \cdots + d = \frac{(d+1)(d+2)}{2}$$

free generators $x_0, x_1, \ldots, x_d$; x_{01}; x_{02}, x_{12}; $x_{0d}, \ldots, x_{d-1,d}$. Further, let $\rho = (V, G)$ be a representation, and let A and B be G-submodules of V such that $A \subset B$ and the quotient B/A is irreducible. Then the stabilizer of this quotient (i.e. the kernel of the representation $(B/A, G)$) is called an *irreducible stabilizer* of ρ.

3.1.4. Lemma. *Let $\rho = (V, G)$ be an arbitrary representation. If $|G| \le d$, then w_d is an identity of ρ. Conversely, if w_d is an identity of ρ, then the index in G of any irreducible stabilizer does not exceed d.*

Proof. If $|G| \le d$, then every substitution of group elements for the variables $x_0, \ldots, x_d$ gives value 1 to at least one of the $x_i^{-1} x_j$. Hence $y \circ w_d \equiv 0$ is satisfied in ρ.

Now, let w_d be an identity of ρ. It suffices to prove that if ρ is a simple representation, then $|G| \le d$. Assume the contrary: $|G| > d$. Choose pairwise distinct elements $g_0, g_1, \ldots, g_d \in G$ and let $0 \ne v \in V$. Since $g_0^{-1} g_1 \ne 1$, Lemma 3.1.3 guarantees that there exists $h_{01} \in G$ such that

$$v_1 = v \circ \left(1 - (g_0^{-1} g_1)^{h_{01}}\right) \ne 0.$$

Hence w_1 is not an identity of ρ. Further, since $g_0^{-1} g_2 \ne 1$ and $g_1^{-1} g_2 \ne 1$, one can successively find $h_{02} \in G$ and $h_{12} \in G$ such that $v_2 = v_1 \circ (1 - (g_0^{-1} g_2)^{h_{02}}) \ne 0$ and $v_3 = v_2 \circ (1 - (g_1^{-1} g_2)^{h_{12}}) \ne 0$. Thus

$$v \circ (1 - (g_0^{-1} g_1)^{h_{02}})(1 - (g_0^{-1} g_2)^{h_{02}})(1 - (g_1^{-1} g_2)^{h_{12}}) \ne 0,$$

that is, w_2 is not an identity of ρ. Repeating this argument, we eventually obtain that w_d is not an identity of ρ, contradicting the assumption. $\square$

Proof of Theorem 3.1.2. Following [44], denote by $\mathcal{C}(e, m, c)$ the class of all groups of exponent e, whose chief factors are of order at most m, and whose nilpotent factors have nilpotency class at most c. By $\mathcal{C}(d; e, m, c)$

we denote the class of all representations $\rho = (V, G)$ over the given field such that $G/\mathrm{Ker}\,\rho \in \mathcal{C}(e, m, c)$ and the indices in G of irreducible stabilizers of ρ are at most d. In particular, if ρ is a simple representation from $\mathcal{C}(d; e, m, c)$, then necessarily $\dim V \leq d$ and $|G| \leq d$. Since any ordinary representation of a finite group belongs to some class $C(d; e, m, c)$ with e not divisible by char K, Theorem 3.1.2 is contained in the following assertion which is of interest in its own right:

if e is not divisible by char K, then $C(d; e, m, c)$ is a Cross variety of group representations.

To prove this assertion, denote $\Theta = \mathcal{C}(e, m, c)$. By a theorem of Kovács and Newman [44], Θ is a Cross variety of groups. Therefore the variety of group representations $\omega\Theta$ is locally finite and finitely based. Denote by $\mathcal{X}$ the variety of all representations from $\omega\Theta$ satisfying the identity w_d. Then $\mathcal{X}$ is also locally finite and finitely based. By Lemma 3.1.4, $\mathcal{X}$ contains only a finite number of simple representations. Since char K does not divide e, all finite representations from $\mathcal{X}$ are ordinary. Thus Lemma 2.5.1 guarantees that $\mathcal{X}$ contains only a finite number of critical representations, and so $\mathcal{X}$ is a Cross variety. It remains to show that $\mathcal{X} = C(d; e, m, c)$.

Evidently $\mathcal{X} \subseteq C(d; e, m, c)$. Conversely, if $\rho \in C(d; e, m, c)$, then ρ certainly belongs to $\omega\Theta$, hence it remains to prove that the identity w_d is satisfied in ρ. It suffices to show that w_d is satisfied in each finitely generated subrepresentation $\sigma = (W, H)$ of ρ. This subrepresentation must be finite, hence, by the Maschke Theorem, it is completely reducible. Let $\sigma = (\bigoplus W_i, H)$ be its decomposition into a sum of H-irreducible summands. Clearly the class $C(d; e, m, c)$ is closed under taking subrepresentations, so that $(W_i, H) \in C(d; e, m, c)$ for each i. Since (W_i, H) is irreducible, it follows that $|H/\mathrm{Ker}\,(W_i, H)| \leq d$. By Lemma 3.1.4, w_d is satisfied in (W_i, H) for each i, hence it is satisfied in σ as well. $\square$

3.1.5. Corollary. *Let $\mathcal{X}$ be a variety of group representations. Then the following conditions are equivalent:*

(a) *$\mathcal{X}$ is generated by an ordinary representation of a finite group;*

(b) *$\mathcal{X} \subseteq C(d; e, m, c)$ where char $K \nmid e$;*

(c) *$\mathcal{X}$ is a Cross variety.*

Indeed, it suffices to notice that every ordinary representation of a finite group is contained in some class $C(d; e, m, c)$ and apply the previous assertion. $\square$

3.2. Several auxiliary results

In the next three sections, our main purpose is to prove that a stable-by-finite representation has a finite basis of identities. To prove this result, we first have to develop the necessary techniques and to establish a number of auxiliary facts, some of which are of independent interest. Our exposition will follow that of [101]; however, it should be noted that the proof incorporates some ideas and intermediate facts from the earlier papers [92] and [78].

Recall that if $\mathcal{X}$ is a variety, then the variety determined by all n-variable identities of $\mathcal{X}$ is denoted by $\mathcal{X}^{(n)}$ (see § 0.3).

3.2.1. Lemma. *If $\mathcal{X}$ is locally finite-dimensional, then $\mathcal{X}^{(n)}$ is finitely based for every n.*

P r o o f. Let $I = \mathrm{Id}\,\mathcal{X}$. Then $\mathcal{X}^{(n)}$ is determined by the set of identities $I_n = I \cap KF_n$. Since $\mathcal{X}$ is locally finite-dimensional, we have

$$\dim_K(KF_n/I_n) < \infty.$$

Thus I_n is an ideal of finite codimension of the finitely generated algebra KF_n, therefore I_n is finitely generated as an ideal, and so $\mathcal{X}^{(n)}$ is finitely based. $\square$

The next assertion generalizes Lemma 3.1.4 and has been proved in [78]. Take the word

$$w_d = w_d\big(x_0, x_1, \ldots, x_d;\, x_{01};\, x_{02}, x_{12};\, \ldots;\, x_{0d}, \ldots, x_{d-1,d}\big)$$

defined in the previous section, and for every k set

$$w_{dk} = w_d\big(x_0, x_1, \ldots, x_d;\, x_{01k};\, x_{02k}, x_{12k};\, \ldots;\, x_{0dk}, \ldots, x_{d-1,d,k}\big)$$

where the x_i and x_{ijk} are, of course, pairwise distinct free generators of the group F. Further, let

$$v_d^{(n)} = w_{d1}w_{d2}\ldots w_{dn}.$$

It is easy to evaluate the number of variables involved in $v_d^{(n)}$: we have $d+1$ variables $x_0, x_1, \ldots, x_d$ plus n sets of $d(d+1)/2$ variables x_{ijk} each, that is, $(d+1) + nd(d+1)/2 = (d+1)(nd+2)/2$ variables in all.

3.2.2. Lemma. *Let $\rho = (V, G)$ be a representation. If G has a normal subgroup H of index $\leq d$ and such that $(V, H) \in \mathcal{S}^n$, then $v_d^{(n)}$ is an identity of ρ. Conversely, if $v_d^{(n)}$ is an identity of ρ, then the index in G of any irreducible stabilizer does not exceed d.*

Proof. To prove the first assertion, let

$$0 = V_0 \subseteq V_1 \subseteq \cdots \subseteq V_n = V$$

be an H-stable series in V. For each $k = 1, \ldots, n$ consider the naturally arising representation $\rho_k = (V_k/V_{k-1}, G)$. Since H acts identically on V_k/V_{k-1}, it follows that $|G/\operatorname{Ker}\rho_k| \leq d$. By 3.1.4, the identity w_{dk} is satisfied in ρ_k. Since this is true for each k, we see that $v_d^{(n)} = w_{d1}\ldots w_{dn}$ is an identity of ρ.

Conversely, let $v_d^{(n)}$ be an identity of ρ. It suffices to show that if ρ is simple, then $|G| \leq d$. Suppose that $|G| > d$ and let $g_0, g_1, \ldots, g_n$ be pairwise distinct elements of G. It was shown in proving Lemma 3.1.4 that for any $0 \neq v \in V$ there exist elements $h_{011}, h_{021}, h_{121}, \ldots, h_{d-1,d,1} \in G$ such that

$$v_1 = v \circ w_{d1}(g_0, g_1, \ldots, g_d; h_{011}; h_{021}, h_{121}; \ldots, h_{d-1,d,1}) \neq 0.$$

Similarly, for v_1 there also exist elements $h_{012}, h_{022}, h_{122}, \ldots, h_{d-1,d,2} \in G$ such that

$$v_2 = v_1 \circ w_{d2}(g_0, g_1, \ldots, g_d; h_{012}; h_{022}, h_{122}; \ldots, h_{d-1,d,2}) \neq 0.$$

Repeating this argument, we eventually obtain that $y \circ v_d^{(n)} \equiv 0$ is not satisfied in ρ, contradicting the assumption. $\square$

The following two properties of stable representations over a field of prime characteristic are commonly known (see for instance [5, 28, 75]).

3.2.3. Lemma. (i) *If (V, G) is a faithful stable representation over a field of characteristic p, then G is a nilpotent p-group of finite exponent and this exponent depends only on p and the class of stability.*

(ii) *If (V, G) is a representation of a finite p-group G over a field of characteristic p, then it is stable of class depending only on $|G|$.* □

For an arbitrary group G, denote by $\Phi(G)$ its Frattini subgroup, by $F(G)$ its Fitting subgroup and by $O_p(G)$ its p-radical (i.e. the largest normal p-subgroup of G). The following assertion follows directly from Lemma 3.2.3.

3.2.4. Corollary. *Let (V, G) be a faithful stable-by-finite representation over a field of characteristic p. Then:*

(i) *$O_p(G)$ is the largest normal subgroup of G acting stably on V;*

(ii) *$O_p(G)$ is the intersection of all irreducible stabilizers of (V, G).* □

For an arbitrary group G, consider its action on $O_p(G)$ via inner automorphisms. Each G-irreducible factor of $O_p(G)$ is an abelian group of exponent p, therefore it can be regarded as a vector space over the prime field $\mathbb{F}_p$. The following lemma is a generalization of one assertion from [78]; the proof belongs to R. Lyons.

3.2.5. Lemma. *Let (V, G) be a finite faithful representation over $\mathbb{F}_p$ and let $O_p = O_p(G)$. If the dimension of each G-irreducible factor of the space V is at most d, then the dimension of each G-irreducible factor of O_p is at most d^2.*

Proof. Let

$$0 = V_0 \subseteq V_1 \subseteq \cdots \subseteq V_t = V$$

be a G-composition series of V. By assumption, $\dim_{\mathbb{F}_p}(V_{j+1}/V_j) \leq d$ for each i. The group G acts on this series in triangular fashion; since this action is faithful, each $g \in G$ can be regarded as a $t \times t$ block-triangular matrix

$$g = \begin{pmatrix} a_{k+1} & & & & & & \\ \ddots & a_{k+2} & & & & \bigcirc & \\ \ddots & \ddots & a_{k+3} & & & & \\ a_4 & \ddots & \ddots & \ddots & & & \\ a_2 & a_5 & \ddots & \ddots & \ddots & & \\ a_1 & a_3 & a_6 & \ddots & a_k & a_{k+t} \end{pmatrix}$$

where $k = 1 + 2 + \cdots + (t-1) = (t-1)t/2$ and each a_i is an $m_i \times n_i$ matrix over $\mathbb{F}_p$ with both m_i and n_i not exceeding d.[1] By the previous assertion, O_p acts trivially on each factor V_{j+1}/V_j, therefore if the above matrix g belongs to O_p, then $a_{k+1} = a_{k+2} = \cdots = a_{k+t} = 1$. In other words, O_p consists of block-unitriangular matrices.

Denote by H_i the set of all matrices g from O_p such that $a_{i+1} = a_{i+2} = \cdots = a_k = 0$ (thus each element from H_i has at most i nonzero element below the main diagonal). Then

$$1 \subset H_1 \subset H_2 \subset \cdots \subset H_k = O_p$$

and it is easy to see that each H_i is a normal subgroup of G. Furthermore, each factor H_i/H_{i-1} can be regarded as an additive $m_i \times n_i$ matrix group over $\mathbb{F}_p$ and so $\dim_{\mathbb{F}_p}(H_{i+1}/H_i) \le d^2$. Since each G-irreducible factor of O_p is isomorphic to a factor of some H_{i+1}/H_i, the lemma follows. $\square$

3.2.6. Lemma. *Let G be a finite group, $O_p = O_p(G)$, $\Phi = \Phi(G)$. Then:*

(i) $O_p(G/\Phi) = O_p\Phi/\Phi$;

(ii) $O_p\Phi/\Phi$ *is a direct product of minimal normal subgroups $M_1/\Phi, \ldots$, M_n/Φ of G/Φ;*

(iii) $O_p\Phi/\Phi$ *has a complement L/Φ in G/Φ, that is, $G/\Phi = (O_p\Phi/\Phi) \rtimes (L/\Phi)$.*

Proof. Denote $F = F(G)$. Then, by a result of Gaschütz [19], we have:

---

[1] We emphasize that the a_i are numbered along the diagonals.

(a) $F(G/\Phi) = F/\Phi$;

(b) F/Φ is a direct product of minimal normal subgroups $M_1/\Phi, \ldots,$ M_r/Φ of G/Φ;

(c) F/Φ has a complement L^*/Φ in G/Φ.

Using (a)–(c) and also the fact that a normal subgroup which is contained in the socle (that is, in the product of *all* minimal normal subgroups) is the product of *some* minimal normal subgroup, we immediately obtain the desired assertions (i)–(iii). □

Our next objective is to prove an analogue of the remarkable "Noncriticality Lemma" of Oates and Powell [69]. To do this, we need Lemmas 0.5.10 and 0.5.11, their group-theoretic prototypes [69, 33.37 and 33.43], and also the following fact.

3.2.7. Lemma (Vovsi [92]). *Let* $\rho = (V, G)$ *be an arbitrary representation,* W *a G-submodule of V and $M_1, \ldots, M_n$ normal subgroups of G. If*

$$V \circ (M_{\pi(1)} - 1)(M_{\pi(2)} - 1)\ldots(M_{\pi(n)} - 1) \subseteq W$$

for every permutation $\pi \in S_n$, *then for every* $\pi \in S_n$

$$V \circ ([M_{\pi(1)}, M_{\pi(2)}, \ldots, M_{\pi(n)}] - 1) \subseteq W.$$

P r o o f. (All commutators without inner brackets are supposed to be left-normed). First we prove the lemma for $n = 2$. In this case, by assumption,

$$V \circ (M_1 - 1)(M_2 - 1) \subseteq W, \quad V \circ (M_2 - 1)(M_1 - 1) \subseteq W. \qquad (1)$$

We will repeatedly use the Three Subgroup Lemma: *if A, B, C are subgroups and N a normal subgroup of some group, and if any two of the commutator subgroups $[A, B, C], [B, C, A], [C, A, B]$ are contained in N, then the third one is contained in N as well.*

Let $V \wr G$ be the semidirect product corresponding to ρ. Then clearly $W \triangleleft V \wr G$. Rewriting (1) in terms of multiplication in the group $V \wr G$, we have

$$[V, M_1, M_2] \subseteq W, \quad [V, M_2, M_1] = [M_2, V, M_1] \subseteq W.$$

By the three subgroup lemma, it follows that $[M_1, M_2, V] \subseteq W$, that is, $[V, [M_1, M_2]] \subseteq W$. In other words, we have got the inclusion

$$V \circ ([M_1, M_2] - 1) \subseteq W,$$

as desired.

Now suppose that everything has been proved for $n - 1$, and let the condition of the lemma be satisfied. One has to show (in multiplicative notation) that $[V, [M_{\pi(1)}, \ldots, M_{\pi(n)}]] \subseteq W$. We will prove this inclusion only when π is the identical permutation, because for any other permutation the proof is analogous. Thus, let us show that $[V, [M_1, \ldots, M_n]] \subseteq W$.

Denote $[M_1, \ldots, M_{n-1}] = M$. In view of the three subgroup lemma, the desired inclusion $[V, [M, M_n]] \subseteq W$ will be established if we prove that

$$[V, M, M_n] \subseteq W, \quad \text{and} \quad [V, M_n, M] \subseteq W. \tag{2}$$

By the condition of the lemma,

$$\forall \pi \in S_{n-1} : \quad [[V, M_{\pi(1)}, \ldots, M_{\pi(n-1)}], M_n] \subseteq W.$$

This means that if W_1/W is the M_n-center of V/W, then

$$\forall \pi \in S_{n-1} : \quad [V, M_{\pi(1)}, \ldots, M_{\pi(n-1)}] \subseteq W_1.$$

Since W_1 is a G-submodule of V, it follows by the induction hypothesis that

$$\forall \pi \in S_{n-1} : \quad [V, [M_{\pi(1)}, \ldots, M_{\pi(n-1)}]] \subseteq W_1.$$

In particular, $[V, [M_1, \ldots, M_{n-1}]] = [V, M] \subseteq W_1$, whence $[V, M, M_n] \subseteq W$. It remains to prove the second of the inclusions (2). By assumption,

$$\forall \pi \in S_{n-1} : \quad [[V, M_n], M_{\pi(1)}, \ldots, M_{\pi(n-1)}] \subseteq W. \tag{3}$$

Denote $V_1 = [V, M]$, then V_1 is a G-submodule of V. Applying the induction hypothesis to (3), we obtain

$$\forall \pi \in S_{n-1} : \quad [V_1, [M_{\pi(1)}, \ldots, M_{\pi(n-1)}]] \subseteq W.$$

In particular, $[V_1, [M_1, \ldots, M_{n-1}]] = [V_1, M] = [V, M_n, M] \subseteq W$, as required. $\square$

3.2.8. Corollary. *Let $\rho = (V, G)$ be a faithful representation, $M_1, \ldots,$ $M_n \triangleleft G$ and*

$$V \circ (M_{\pi(1)} - 1) \ldots (M_{\pi(n)} - 1) = 0$$

for every $\pi \in S_n$. Then for every $\pi \in S_n$

$$[M_{\pi(1)}, \ldots, M_{\pi(n)}] = 1. \quad \square$$

In particular, for $M_1 = \cdots = M_n = G$ this yields the well known Kaloujnine theorem [38]: if (V, G) is a faithful n-stable representation, then G is a nilpotent group of class at most $n - 1$.

We can now establish the desired Noncriticality Lemma. It was proved in [92] and is an analogue of Lemma 2.4.2 from Oates and Powell [69] (see also [68, 51.37]).

3.2.9. Lemma. *Let $\rho = (V, G)$ be a faithful cyclic representation. Suppose that G possesses normal subgroups $M_1, \ldots, M_n$ and a subgroup L such that*

(a) $G = \langle M_1, \ldots, M_n, L \rangle$;

(b) $G \neq \langle M_1, \ldots, M_{i-1}, M_{i+1}, \ldots, M_n, L \rangle$ $(i = 1, \ldots, n)$;

(c) $V \circ (M_{\pi(1)} - 1) \ldots (M_{\pi(n)} - 1) = 0$ *for every permutation $\pi \in S_n$.*

Then ρ is not critical.

P r o o f. Let $H_i \simeq M_i$ $(i = 1, \ldots, n)$ and $H_0 \simeq L$. Consider the free product $H = H_0 * H_1 * \cdots * H_n$ and apply Lemmas 0.5.10 and 0.5.11 to the regular representation (KH, H). For each $i = 1, \ldots, n$ let θ_i be the endomorphism of (KH, H) taking H_i to 1 and acting trivially on each H_j $(j \neq i)$. Denote

$$(A_i, D_i) = \operatorname{Ker} \theta_i, \quad (A, D) = \bigcap_{i=1}^{n} (A_i, D_i).$$

By 0.5.10, each element of A is a linear combination of monomials involving elements from every H_i. By [68, 33.37], each element of D is a product of commutators also involving elements from every H_i.

Let α be the epimorphism of H onto G extending the given isomorphisms $H_0 \simeq L$ and $H_i \simeq M_i$. If v is a generator of the cyclic KG-module V then, letting $1_H^\alpha = v$, we get an epimorphism of representations $\alpha : (KH, H) \to \rho$. It follows from (c) and Corollary 3.2.8 that

$$[M_{\pi(1)}, M_{\pi(2)}, \ldots, M_{\pi(n)}] = 1$$

for every $\pi \in S_n$. Using this observation together with (c) and taking into account the structure of elements from A and D described in the previous paragraph, we see that $(A, D) \subseteq \operatorname{Ker}\alpha$.

Let J be the set of all nonempty sequences $j = (j_1, \ldots, j_r)$, where $1 \leq j_1 < \cdots < j_r \leq n$. By Lemma 0.5.11, every $w \in KH$ can be written as

$$w = u + \sum_{j \in J} \pm w\beta_j$$

where $u \in A \subseteq \operatorname{Ker}\alpha$ and $\beta_j = \theta_{j_1} \ldots \theta_{j_r}$. A similar decomposition is valid for any element of the group H [68, 33.43]. In particular, this guarantees that if we define $(V_j, N_j) = (KH, H)^{\beta_j \alpha}$, then (V_j, N_j) is contained in some subrepresentation (V, G_i) of ρ, where $G_i = \langle M_1, \ldots, M_{i-1}, M_{i+1}, \ldots, M_n, L \rangle$. By (b), (V_j, N_j) is a proper subrepresentation of ρ.

Let ϕ be the direct product of the representations (V_j, N_j):

$$\phi = \prod_{j \in J}(V_j, N_j) = \left(\bigoplus_{j \in J} V_j, \prod_{j \in J} N_j \right).$$

Using the epimorphisms $\beta_j \alpha : (KH, H) \to (V_j, N_j)$, define a homomorphism $\gamma : (KH, H) \to \phi$ letting for any $w \in KH$ and $h \in H$

$$w\gamma = \sum_{j \in J} w\beta_j\alpha, \quad h\gamma = \prod_{j \in J} h\beta_j\alpha. \tag{4}$$

Let us show that $\operatorname{Ker}\gamma \subseteq \operatorname{Ker}\alpha$.

Denote $\operatorname{Ker}\gamma = (B_\gamma, R_\gamma)$, $\operatorname{Ker}\alpha = (B_\alpha, R_\alpha)$. The inclusion $R_\gamma \subseteq R_\alpha$ was proved in the original noncriticality lemma. We prove that $B_\gamma \subseteq B_\alpha$. According to (4), the equality $w\gamma = 0$ is valid in the direct sum $\bigoplus V_j$ if and only if $w\beta_j\alpha = 0$ for all j. In other words, $w\beta_j \in \operatorname{Ker}\alpha$ for all j. Since $w = u + \sum \pm w\beta_j$ and $u \in \operatorname{Ker}\alpha$, it follows that $w \in \operatorname{Ker}\alpha$.

Thus, we have the epimorphism $\alpha : (KH, H) \to \rho$ and the homomorphism $\gamma : (KH, H) \to \phi$ such that $\operatorname{Ker} \gamma \subseteq \operatorname{Ker} \alpha$. Therefore there exists an epimorphism of $(KH, H)\gamma$ onto ρ. Since $(KH, H)\gamma$ is a subrepresentation of ϕ and ϕ is a direct product of proper subrepresentations (V_j, N_j) of ρ, we see that ρ cannot be critical. $\square$

3.3. Bounds for the orders of critical representations

The aim of this section is to show that, under certain conditions, the orders of critical representations and the cardinalities of their generating sets are bounded by several special parameters. These results will play the key role in proving the finite basis property for stable-by-finite representations.

3.3.1. Lemma. *Let $\rho = (V, G)$ be a critical representation, $V^* = \mu(\rho)$ its monolith and N a nilpotent normal subgroup of G acting trivially on V^*. Then the representation (V, N) is stable, and if also $\operatorname{char} K \nmid |N|$, then this representation is trivial.*

Proof. Let $N = N_p \times N_q \times \ldots$ be the decomposition of N into primary components. Each N_p with $p \nmid \operatorname{char} K$ acts on V completely reducibly. Let

$$V = V_1 \oplus V_2 \oplus \cdots \oplus V_n$$

be the decomposition of V into N_p-irreducible summands. Let A be the sum of all V_i on which N_p acts trivially, and let B be the sum of all other V_i. Then $V = A \oplus B$. Since N_p acts trivially on V^*, we have necessarily $A \neq \{0\}$, and since $N_p \lhd G$, both A and B are invariant with respect to G. However, V is an indecomposable G-module because $\rho = (V, G)$ is critical. Hence $V = A$, that is, N_p acts trivially on V for every $p \neq \operatorname{char} K$. It remains to note that if $p = \operatorname{char} K$, then N_p acts stably on V by Lemma 3.2.3 (ii). $\square$

Until the end of this section we assume that $\operatorname{char} K = p \neq 0$. In addition, we fix the following notation:

$$\rho = (V, G) \text{ is a critical representation over } K,$$

$$\Phi = \Phi(G), \quad O_p = O_p(G), \quad \Phi_p = \Phi(O_p),$$

$$C = C_G(O_p\Phi/\Phi), \quad C_p = C_G(O_p/\Phi_p).$$

It is well known that $A \lhd B \implies \Phi(A) \subseteq \Phi(B)$. Therefore $\Phi_p \subseteq \Phi$, whence $C_p \subseteq C$. Moreover, since O_p/Φ_p is an abelian group of exponent p, we have $O_p \subseteq C_p$. Thus

$$\Phi_p \subseteq O_p \subseteq C_p \subseteq C \subseteq G \tag{1}$$

where all the terms of this chain are normal in G. Finally, for any stable representation, we agree for brevity to call its stability class just the *class* of this representation.

3.3.2. Lemma. *Suppose that the classes of stable quotients of the critical representation $\rho = (V, G)$ are at most s, while the indices in G of irreducible stabilizers of this representation are at most d. Then $|C/O_p| \leq d^{sd+1}$.*

P r o o f. 1) By Lemma 3.2.4 (ii), O_p coincides with the intersection of all the irreducible stabilizers C_j of ρ. Since $O_p \subseteq C$, we have $O_p = \bigcap_j (C \cap C_j)$. Let $V^* = \mu(V, G)$ and let C^* be the kernel of the representation (V^*, G). Then C^* is one of the irreducible stabilizers of ρ, and hence $O_p = \bigcap_j (C \cap C^* \cap C_j)$. Dropping the extra terms from this intersection, we obtain an "irredundant" intersection

$$O_p = \bigcap_{j=1}^{t} (C \cap C^* \cap C_j). \tag{2}$$

So $O_p = (C \cap C_1) \cap \cdots \cap (C \cap C_t) \cap (C \cap C^*)$, and hence C/O_p can be embedded isomorphically in the direct product $G/C_1 \times \cdots \times G/C_t \times G/C^*$. Since all of $C_1, \ldots, C_t, C^*$ are irreducible stabilizers, the order of every factor here is at most d. Hence $|C/O_p| \leq d^{t+1}$, and it remains only to show that $t \leq sd$.

2) We can assume that $t > d$, for otherwise there is nothing to prove. We put

$$D_i = \bigcap_{j \neq i} (C \cap C^* \cap C_j), \quad i = 1, \ldots, t.$$

Then $D_i \cap D_k = O_p$ for $i \neq k$; that is, D_i and D_k centralize each other modulo O_p. Note that the groups D_i/O_p are nontrivial by the minimality of the intersection (2). A standard argument shows that they generate their direct product in G/O_p. Indeed, let $\prod d_i \in O_p$. Since the factors of this product commute modulo O_p, we have

$$d_j^{-1} \equiv \prod_{i \neq j} d_i \mod O_p, \quad (j = 1, \ldots, t).$$

Each factor on the right hand side belongs to C_j by the definition of D_i, while $d_j \in D_j$. Since $C_j \cap D_j = O_p$, we conclude that $d_j \in O_p$. Thus every factor of the product $\prod d_i$ belongs to O_p, whence the desired assertion follows.

In each D_i/O_p we choose a subgroup M_i/O_p which is a minimal non-trivial normal subgroup of G/O_p. Suppose that the group $M_i\Phi/\Phi$ is nilpotent. Then by (a) from the proof of Lemma 3.2.6, $M_i \subseteq F(G)$. Since $M_i \subseteq D_i \subseteq C^* = \mathrm{Ker}\,(V^*, G)$, it follows from Lemma 3.3.1 that M_i acts stably on V. So M_i is a p-group, and hence $M_i = O_p$. But this is impossible, since M_i/O_p is nontrivial. Therefore for every i the group $M_i\Phi/\Phi$ is not nilpotent.

Suppose now that M_i/O_p is a nilpotent group. Since $M_i \subseteq C$ and C centralizes the factor $O_p\Phi/\Phi$, which is itself a nilpotent group, the series $M_i\Phi \supseteq O_p\Phi \supseteq \Phi$ can be refined into a central series, and hence the group $M_i\Phi/\Phi$ is nilpotent. By the above, this is impossible, and hence M_i/O_p is not nilpotent. Thus we have proved that there are non-nilpotent normal subgroups M_i/O_p of G/O_p, generating their direct product in the latter. Recall now a well known group-theoretic fact [68, 52.43]:

If $M_1, \ldots, M_n$ are non-nilpotent normal subgroups of a finite group G which generate their direct product in G, then G possesses a subgroup L such that $G = \langle M_1, \ldots, M_n, L \rangle$, but G is not generated by L together with a proper subset of the set of subgroups $M_1, \ldots, M_n$.

Applying this statement to our group G/O_p, we conclude that G/O_p has a subgroup L/O_p such that

$$G/O_p = (M_1/O_p \times \cdots \times M_t/O_p) \cdot L/O_p, \tag{3}$$

where no M_i/O_p can be removed without violating the equality (3). Then $G = M_1 \ldots M_t L$, and no subgroups M_i can be removed. Thus the representation $\rho = (V, G)$ and the subgroups $M_1, \ldots, M_t, L$ satisfy the first two conditions of the noncriticality lemma. We show that if t is sufficiently large, then they also satisfy the third condition of the lemma.

3) Since the group M_i/O_p is not nilpotent, it has elements of order prime to p. Being a chief factor of G, M_i/O_p is a direct power of some simple group. Hence M_i/O_p is generated by its elements of order prime to p. Furthermore, let

$$0 = V_0 \subset V_1 \subset \cdots \subset V_m = V \tag{4}$$

be an upper O_p-stable series in V. We fix an arbitrary factor V_k/V_{k-1} and refine it to a G-composition series

$$V_{k-1} = W_0 \subset W_1 \subset \cdots \subset W_l = V_k.$$

Then we choose elements $b_1, \ldots, b_d$ in G (where d is the bound for the indices of irreducible stabilizers given in the statement of the lemma) which satisfy the following conditions:

a) Every b_j belongs to some subgroup M_{i_j} $(1 \le i_j \le t)$.

b) Distinct b_j belong to distinct subgroups M_{i_j} (this is possible since, by assumption, $t \ge d$).

c) The order of every b_j modulo O_p is prime to p.

We put $B = \langle O_p, b_1, \ldots, b_d \rangle$ and study the representation $(V_k/V_{k-1}, B)$. The group B acts on V_k/V_{k-1} as B/O_p. The group B/O_p is abelian (since distinct M_i commute elementwise modulo O_p) and is generated by elements of order prime to p. Therefore $p \nmid |B/O_p|$, whence $(V_k/V_{k-1}, B)$ is completely reducible. Consequently, each W_{r-1}/V_{k-1}, where $r = 1, \ldots, l$, has a B-invariant direct complement in W_r/V_{k-1}, which we denote by U_{r-1}/V_{k-1}. So $W_r/V_{k-1} = W_{r-1}/V_{k-1} \oplus U_{r-1}/V_{k-1}$ and

$$V_k/V_{k-1} = \bigoplus_{r=1}^{l} U_{r-1}/V_{k-1}. \tag{5}$$

We show that $U_{r-1} \circ (b_1 - 1) \ldots (b_d - 1) \subseteq V_{k-1}$ or, equivalently, that

$$W_r \circ (b_1 - 1) \ldots (b_d - 1) \subseteq W_{r-1}.$$

For this it suffices to show that at least one of the b_j lies in the stabilizer of the quotient W_r/W_{r-1}. We put $C_r = C_G(W_r/W_{r-1})$ and consider the group

$$C_r/O_p \cap (M_1/O_p \times \cdots \times M_t/O_p).$$

It is contained in the socle of the group G/O_p and is normal in G/O_p, therefore it is the direct product of some of the M_i/O_p. Since C_r is an irreducible stabilizer of ρ, we have $|G/C_r| \leq d$. Hence there are at most $d-1$ of the M_i/O_p not contained in C_r/O_p, whence at least one of the groups M_{i_j}/O_p $(j = 1,\ldots,d)$ is contained in C_r/O_p. Then the corresponding b_j belongs to C_r, as required.

Thus on all the direct summands U_{r-1}/V_{k-1} from (5) the element $(b_1 - 1)\ldots(b_d - 1)$ acts as zero. Hence

$$V_k \circ (b_1 - 1)\ldots(b_d - 1) \subseteq V_{k-1}. \tag{6}$$

We emphasize that this is true for all $k = 1,\ldots,m$ and for all sets of elements $b_1,\ldots,b_d$ satisfying the conditions a)–c).

4) We now show that if $t \geq sd + 1$, then

$$V \circ (M_{\pi(1)} - 1)\ldots(M_{\pi(t)} - 1) = 0$$

for any permutation π of the numbers $1,\ldots,t$. Let $x_i \in M_{\pi(i)}$; we must show that $V \circ (x_1 - 1)\ldots(x_t - 1) = 0$. Since $s \geq m$ (we recall that s is the bound for the stability classes of the factors of the representation ρ, while m is the length of the series (4)), we have $t \geq sd + 1 \geq md + 1 > md$, so it suffices to check that

$$V_k \circ (x_1 - 1)\ldots(x_d - 1) \subseteq V_{k-1}. \tag{7}$$

Since the groups M_i/O_p are generated by their elements of order prime to p, every x_i can be presented as $x_i = b_1\ldots b_n a$, where $b_\alpha \in M_{\pi(i)}$, $a \in O_p$ and the orders of the elements b_α modulo O_p are not divisible by p. The desired inclusion is now immediately deduced from (6), the formula

$$gh - 1 = (g - 1)(h - 1) + (g - 1) + (h - 1),$$

and the fact that O_p acts identically on V_k/V_{k-1} and that distinct M_i commute elementwise modulo O_p.

Thus, for $t \geq sd + 1$, the representation ρ satisfies all the conditions of the noncriticality lemma, which is impossible. Hence $t \leq sd$, and so $|C/O_p| \leq d^{sd+1}$. □

3.3.3. Lemma. *Let the classes of stable factors of a critical representation* $\rho = (V, G)$ *be at most* s, *the indices of irreducible stabilizers at most* d, *and the orders of chief factors of the group* G *at most* m. *Then the number of generators of* G *is bounded by a number depending only on* s, d *and* m, *and the order of* G *is bounded by a number depending only on* s, d, m *and* ρ.

P r o o f. We adhere to the notation of Lemma 3.2.6. We have

$$O_p \Phi / \Phi = M_1 / \Phi \times \cdots \times M_r / \Phi$$

where all the M_i / Φ are chief factors of G. Then

$$C = C_G(O_p \Phi / \Phi) = \bigcap_{i=1}^{r} C_G(M_i / \Phi)$$

and since $|M_i / \Phi| \leq m$, we have $|G / C_G(M_i / \Phi)| \leq m!$ and hence $|G/C| \leq (m!)^r$.

In each of the normal subgroups M_i we choose a Sylow p-subgroup P_i $(i = 1, \ldots, r)$. Since M_i is nilpotent, $M_i = P_i \times Q_i$, where $p \nmid |Q_i|$; therefore from the fact that M_i / Φ is a p-group, we have $Q_i \subseteq \Phi$. Hence $M_i = P_i \Phi$, and since

$$G / \Phi = (M_1 / \Phi \times \cdots \times M_r / \Phi) \cdot L / \Phi,$$

we have $G = \langle M_1, \ldots, M_r, L \rangle = \langle P_1, \ldots, P_r, L \rangle$, where none of the subgroups can be omitted. The representation (V, O_p) is stable of class at most s, while all the P_i are contained in O_p. Thus, if $r \geq s$, all the conditions of the noncriticality lemma are satisfied by the subgroups $P_1, \ldots, P_r$ and L. This is impossible, and so $r < s$ whence $|G/C| < (m!)^s$.

By Lemma 3.3.2, $|C/O_p| \leq d^{sd+1}$, and therefore $|G/O_p| < d^{sd+1}(m!)^s$. It remains to bound both the order of O_p and its number of generators.

For brevity, we denote $|G/O_p|$ by k, and suppose that R is the subgroup generated by the complete set of coset representatives of the factor group G/O_p. The action of G by conjugation on O_p induces a faithful representation $(O_p / \Phi_p, G / C_p)$ over $\mathbb{F}_p$ (we recall that $\Phi_p = \Phi(O_p)$

.

and $C_p = C_G(O_p/\Phi_p)$). We choose in O_p/Φ_p a set of cyclic $\mathbb{F}_p[G/C_p]$-submodules $N_1/\Phi_p, \ldots, N_t/\Phi_p$ in such a way that $G = N_1 \ldots N_t R$ with t minimal. Then by the noncriticality lemma we get $t \leq s$.

Since every N_i/Φ_p is a cyclic $\mathbb{F}_p[G/C_p]$-module, we have

$$\dim_{\mathbb{F}_p}(N_i/\Phi_p) \leq |G/C_p| \leq |G/O_p| = k.$$

Consequently, each group N_i/Φ_p has at most k generators, and hence

$$G/\Phi_p = N_1 \ldots N_t R/\Phi_p$$

has at most $k(s+1)$ generators. Since $\Phi_p \subseteq \Phi$, the group G/Φ has at most $k(s+1)$ generators as well. Using the well known property of the Frattini subgroup, we conclude that the number of generators of G is also bounded by

$$k(s+1) < d^{sd+1}(m!)^s(s+1).$$

Finally, the bounds for the number of generators of G and for the index of O_p in G give a bound for the number of generators of O_p (by the Schreier formula). Since O_p acts faithfully and s-stably on V, O_p is an $(s-1)$-nilpotent p-group whose exponent is bounded in terms of s and p (Lemma 3.2.3 (i)). Together with the above, this gives a bound for $|O_p|$, and hence also for $|G|$. $\square$

3.3.4. Lemma. *Let the classes of stable factors of a critical representation $\rho = (V, G)$ be at most s, the indices of irreducible stabilizers at most d, and let $e = \exp(G/O_p)$ be not divisible by p. Then the number of generators of G is bounded by a number depending only on s, d and e, and the order of G is bounded by a number depending only on s, d, e and p.*

Proof. As in the preceding lemma, we consider the representation $(O_p/\Phi_p, G/C_p)$ over $\mathbb{F}_p$. Since $O_p \subseteq C_p$, it follows that $p \nmid |G/C_p|$, and hence this representation is completely reducible. Let

$$O_p/\Phi_p = M_1/\Phi_p \oplus \cdots \oplus M_r/\Phi_p$$

be a decomposition of O_p/Φ_p into $\mathbb{F}_pG$-irreducible summands. Since O_p is a normal subgroup of G whose index is prime to p, we have $G = O_p \rtimes L$ for some subgroup L. Hence $G = \langle M_1, \ldots, M_r, L \rangle$, where all M_i are normal in G and no M_i can be omitted. By the noncriticality lemma, $r < s$.

Being a representation over the field K, ρ is also a representation over its prime subfield $\mathbb{F}_p$. By the condition, the index in G of the stabilizer of each irreducible factor of the KG-module V is at most d. But then the index in G of the stabilizer of each irreducible factor of the $\mathbb{F}_p G$-module V is also at most d (it is clear that every stabilizer of a $\mathbb{F}_p G$-irreducible factor contains the stabilizer of some KG-irreducible factor). Therefore the $\mathbb{F}_p$-dimension of each $\mathbb{F}_p G$-irreducible factor of the module V is at most d. By Lemma 3.2.5, the $\mathbb{F}_p$-dimension of each $\mathbb{F}_p G$-irreducible factor of O_p is at most d^2. Thus $\dim_{\mathbb{F}_p}(M_i/\Phi_p) \le d^2$, whence

$$\dim_{\mathbb{F}_p}(O_p/\Phi_p) \le d^2 s. \tag{8}$$

Therefore, first, it follows that G/C_p is a matrix group over $\mathbb{F}_p$ of degree at most $d^2 s$ and of exponent dividing e. By the classical Burnside Theorem, $|G/C_p|$ is bounded by a number $f_1(d^2 s, e)$. Second, it follows from (8) that the group O_p/Φ_p has at most $d^2 s$ generators, and hence O_p itself can be generated by at most $d^2 s$ elements. As in the preceding lemma, it now follows that $|O_p| \le f_2(d, s, p)$.

We have $O_p \subseteq C_p \subseteq G$. The orders of the groups G/C_p and O_p, and the number of generators for O_p have already been bounded. In addition, by (1) and Lemma 3.3.2 we have

$$|C_p/O_p| \le |C/O_p| \le d^{sd+1}.$$

Thus the number of generators of the group G is at most $f_1(d^2 s, e) + d^{sd+1} + d^2 s$, while the order of this group is at most $f_1(d^2 s, e) d^{sd+1} f_2(d, s, p)$. $\square$

Remark. We emphasize that in Lemmas 3.3.3 and 3.3.4 the number of generators for the group G (in contrast to its order!) is independent of the ground field K. This observation will be of decisive importance in the proof of the main theorem over a field of characteristic zero.

3.4. Identities of stable-by-finite representations

We say that a variety $\mathcal{X}$ of group representations is a *variety of bounded rank* if it satisfies the following conditions:

(i) $\mathcal{X}$ is locally finite-dimensional,

(ii) $\mathcal{X}$ is finitely based,

(iii) the basis ranks of all its subvarieties (including $\mathcal{X}$) are finite and uniformly bounded.

3.4.1. Lemma. *All subvarieties of a variety of bounded rank are also varieties of bounded rank and satisfy the ascending and the descending chain conditions.*

Proof. Let $\mathcal{X}$ be a variety of bounded rank, and let n be an upper bound for the basis ranks of its subvarieties. Then $\mathcal{X}$ is generated by its free representation $\mathrm{Fr}_n\mathcal{X} = (KF_n/I_n, F_n)$ of rank n (here I_n is the ideal of identities of $\mathcal{X}$ in the group algebra KF_n). Since all subvarieties of $\mathcal{X}$ are generated by group representations with n generators, it is easy to see that all these subvarieties are in one-to-one correspondence with the verbal overideals of I_n in the algebra KF_n. By the local finite-dimensionality of $\mathcal{X}$, the algebra KF_n/I_n is finite-dimensional, and hence the subvarieties of $\mathcal{X}$ satisfy both chain conditions. The rest is obvious. $\square$

3.4.2. Corollary. *Every variety of bounded rank is Specht, and every Cross variety is a variety of bounded rank.* $\square$

Thus the long preparatory work has been finished, and now we can state and prove the main result of the present chapter.

3.4.3. Theorem (Vovsi and Nguyen Hung Shon [101]). *The variety generated by a stable-by-finite representation is a variety of bounded rank.*

The proof of this theorem divides into two cases, depending on the characteristic of the ground field.

THE CASE OF PRIME CHARACTERISTIC. Let $\operatorname{char} K = p$. Recall that $\mathcal{C}(e, m, c)$ denotes the class of all groups of exponent e, whose chief factors have orders at most m, and whose nilpotent factors have classes at most c. According to [44], $\mathcal{C}(e, m, c)$ is a Cross variety of groups. By $\mathcal{C}(d, s; e, m, c)$ we denote the class of all representations $\rho = (V, G)$ over K such that $G/\mathrm{Ker}\,\rho \in \mathcal{C}(e, m, c)$, the indices in G of irreducible stabilizers of ρ are at most d, and the classes of stable factors of ρ are at most s (cf. the proof of Theorem 3.1.2).

3.4.4. Lemma. *Let $\rho = (V, G)$ be an stable-by-finite representation, and let $\mathcal{X} = \mathrm{var}\,\rho$. Then there are positive integers k, d, s, e, m, c for which $\mathcal{X}^{(k)} \subseteq \mathcal{C}(d, s; e, m, c)$.*

P r o o f. 1) By hypothesis, there is a normal subgroup H of G such that $(V, H) \in \mathcal{S}^n$ and $|G/H| = d < \infty$. By Lemma 3.2.2, ρ satisfies the identity $v_d^{(n)}$ depending on $(d+1)(dn+2)/2 = r$ variables. Hence this identity also holds in the variety $\mathcal{X}$ and in the variety $\mathcal{X}^{(r)}$. Applying Lemma 3.2.2 once more, we find that the indices of irreducible stabilizers in any representation from $\mathcal{X}^{(r)}$ are at most d.

2) Without loss of generality, the representation ρ will be assumed to be faithful. Then, by Lemma 3.2.3 (i), H is a nilpotent p-group of finite exponent. Consequently every Sylow p-subgroup of G contains H. Let

$$0 = A_0 \subset A_1 \subset \cdots \subset A_l = V \qquad (l \leq n) \qquad (1)$$

be the upper H-stable series in V. If P is an arbitrary Sylow p-subgroup of G, then $H \lhd P$, and hence all terms in (1) are invariant under P. On the factors A_{i+1}/A_i the group P acts as P/H. Since P/H is a finite p-group of order at most d, by Lemma 3.2.3 (ii) every representation $(A_{i+1}/A_i, P/H)$ is stable of class depending only on d. Therefore (1) can be refined to a P-stable series whose length is bounded by a constant depending only on d and n. We denote this constant by $s = s(d, n)$. Thus if P is a Sylow p-subgroup of G, then the representation (V, P) is stable of class at most s.

Next we show that all stable representations of the variety $\mathcal{X}^{(s)}$ have stability class at most s. First, let (A, N) be a finite faithful stable representation in $\mathcal{X}$. Then N is a finite p-group. Applying Lemma 2.1.5 and taking into account that (A, N) is faithful, we see that $(A, N) \in \mathrm{QSD}_0\{(V, G)\}$.

This means that there is an epimorphism $\alpha : (B,S) \to (A,N)$, where $(B,S) \subseteq (V_1,G_1) \times \cdots \times (V_t,G_t)$ and all the (V_i,G_i) are isomorphic copies of the initial $\rho = (V,G)$. Since (A,N) is finite, (B,S) can be replaced by a finitely generated subrepresentation. By 2.2.2 and 2.2.9, $\mathcal{X}$ is a locally finite variety; hence this subrepresentation is finite. Let (B,S) be already finite. Since N is a finite p-group, it is an α-epimorphic image of some Sylow p-subgroup P of S. Thus we have an epimorphism

$$\alpha : (B,P) \to (A,N) \quad \text{where} \quad (B,P) \subseteq (V_1,G_1) \times \cdots \times (V_t,G_t).$$

It is clear that then $(B,P) \subseteq (V_1,P_1) \times \cdots \times (V_t,P_t)$, where P_i is some Sylow p-subgroup of G_i. From the above it now follows that the stability class of the representation (B,P) is at most s. A fortiori this is true for (A,N).

Now let (A,N) be an arbitrary stable representation in $\mathcal{X}^{(s)}$. Then all its $(1,s)$-generated subrepresentations belong to $\mathcal{X}$. These subrepresentations are finite and, as has just been shown, they all belong to S^s. Therefore $(A,N) \in (S^s)^{(s)}$. It remains to note that, by Corollary 1.1.2, $(S^s)^{(s)} = S^s$.

3) We show that the group G belongs to the class $\mathcal{C}(e,m,c)$ for some e, m and c. For e we can immediately take the exponent of G. Furthermore, let A be a nilpotent factor of G; then A is an extension of a p-group B of a finite exponent by a finite group A/B, where

$$\exp B \le \exp H \le e \quad \text{and} \quad |A/B| \le |G/H| = d.$$

Being a nilpotent group, A can be presented as $A = P \times Q_1 \times Q_2 \times \ldots$, where P is its Sylow p-subgroup and the Q_i are Sylow q_i-subgroups ($q_i \ne p$). Then $B \subseteq P$, and thus P is an extension of the p-group B of exponent at most e by the finite p-group P/B of order at most d. For such a group, its nilpotence class is bounded by some $f(e,d)$ (see for example [5]). As for an arbitrary Q_i, we have $Q_i \cap B = 1$, whence $|Q_i| \le d$ and so the nilpotence class of Q_i can not exceed d. Hence the nilpotence class of A is also bounded by some $c = c(d,e)$.

It remains to bound the orders of the chief factors of G by some number m. By the structure of G, it follows that each of its chief factors is either a factor of G/H (and so has order at most d) or is a central factor of the group H of exponent p. In the second case, this factor is actually an irreducible $\mathbb{F}_p[G/H]$-module. The order of such a module is at most p^d, and hence it suffices to put $m = p^d$. Thus $G \in \mathcal{C}(e, m, c)$ for some e, m and c.

4) Set $\Theta = \mathcal{C}(e, m, c)$, then $\mathcal{X} = \mathrm{var}\,\rho \subseteq \omega\Theta$. Since Θ is a Cross variety of groups, $\omega\Theta$ is finitely based. A fortiori, $\omega\Theta$ has a finite axiomatic rank, that is, $(\omega\Theta)^{(t)} = \omega\Theta$ for some t. Hence $\mathcal{X}^{(t)} \subseteq \omega\Theta$.

Finally we put $k = \max(r, s, t)$, where the numbers r, s and t are as indicated in parts 1), 2) and in the preceding paragraph respectively. By the above it follows that $\mathcal{X}^{(k)} \subseteq \mathcal{C}(d, s; e, m, c)$. $\square$

The proof of Theorem 3.4.3 in the case $\mathrm{char}\,K = p$ can now be completed quickly. Let ρ be a stable-by-finite representation over K, and let $\mathcal{X} = \mathrm{var}\,\rho$. By the preceding lemma, $\mathcal{X}^{(k)} \subseteq \mathcal{C}(d, s; e, m, c)$ for some k, d, s, e, m, c. Therefore $\mathcal{X}^{(k)} \subseteq \omega\Theta$, where $\Theta = \mathcal{C}(e, m, c)$. By Theorem 2.2.2, the variety $\mathcal{X}^{(k)}$ is locally finite, and by Lemma 3.2.1 it is finitely based. All subvarieties of $\mathcal{X}^{(k)}$ are generated by their critical representations, but according to Lemma 3.3.3, if (V, G) is a critical representation in $\mathcal{C}(d, s; e, m, c)$, then the number of generators for G can be bounded by some $r = r(d, s, e)$. Consequently the basis ranks of all subvarieties of $\mathcal{X}^{(k)}$ are at most r. Thus $\mathcal{X}^{(k)}$ is a variety of bounded rank. The same holds for its subvariety $\mathcal{X}$.

THE CASE OF ZERO CHARACTERISTIC. 1) Now let $\mathrm{char}\,K = 0$, and let $\rho = (V, G)$ be a stable-by-finite representation over K. The group G possesses a normal subgroup H such that $(V, H) \in \mathcal{S}^n$ and $|G/H| = d < \infty$. By Lemma 3.2.2, ρ satisfies the identity $v_d^{(n)}$. Let $\Theta = \mathrm{var}(G/H)$ and let $\mathcal{V}$ be the variety of all group representations from $\mathcal{S}^n \times \Theta$ satisfying the identity $v_d^{(n)}$. Since $\mathrm{var}\,\rho \subseteq \mathcal{V}$, it suffices to show that $\mathcal{V}$ is a variety of bounded rank. By Theorem 2.2.1, $\mathcal{V}$ is locally finite-dimensional. By the Oates–Powell theorem, Θ is a Cross variety of groups; therefore, by Theorem 1.1.4, the variety $\mathcal{S}^n \times \Theta$ is finitely based. Hence $\mathcal{V}$ is finitely based as well. It remains to prove that if $\mathcal{Y}$ is a subvariety of $\mathcal{V}$, then $r_b(\mathcal{Y})$ is bounded by some $r = r(\mathcal{V})$.

2) Let R be a finitely generated subring of K. We consider the map $\nu' : \mathrm{M}(K) \to \mathrm{M}(R)$ defined in §0.4, and we put $\mathcal{Y}^{\nu'} = \mathcal{X}$. Obviously $\mathcal{X} \subseteq (\mathcal{S}^n \times \Theta)_R$ (the subscript R shows that the corresponding variety of representations is being considered over R). Denote by π the set of all primes not dividing $e = \exp \Theta$. Since $\mathcal{Y}$ is locally finite-dimensional, by Proposition 0.4.5 the variety $\mathcal{X}$ is generated by finite R-representations over finite fields with characteristics in π. Hence $\mathcal{X}$ is also generated by critical R-representations over finite fields with characteristics in π. More precisely, $\mathcal{X}$ is generated by representations (A, N) over R, such that:

(a) $R/\mathrm{Ann}_R A = L$ is a finite field and $\mathrm{char}\, L \nmid e$;

(b) (A, N), regarded as a representation over L, is critical.

Our next objective is to show that for each representation (A, N) of this sort the number of generators for the group N is bounded in terms of d, n and e.

3) Denote by $\mathcal{X}_L$ the class of all representations from $\mathcal{X}$ for which the domain of action is annihilated by the ideal $\mathrm{Ann}_R A$. It is clear that $\mathcal{X}_L$ is actually a variety of group representations over L, and immediately from the definition we have $(A, N) \in \mathcal{X}_L$.

Since the identity $v_d^{(n)}$ is satisfied in $\mathcal{Y}$ and all its coefficients are integers, it is also satisfied in both $\mathcal{X} = \mathcal{Y}^{\nu'}$ and $\mathcal{X}_L$. By Lemma 3.2.2, the indices of irreducible stabilizers in any representation from $\mathcal{X}_L$ (in particular in (A, N)) are at most d. Furthermore, let $\mathrm{char}\, L = p$. Since

$$(A, N) \in \mathcal{X}_L \subseteq \mathcal{S}_L^n \times \Theta,$$

we have $(A, \Theta^*(N)) \in \mathcal{S}_L^n$. Hence, from the faithfulness of the critical representation (A, N) we conclude that $\Theta^*(N)$ is a p-group, and so $\Theta^*(N) \subseteq O_p(N) = O_p$. Actually, $\Theta^*(N) = O_p$, because $e = \exp \Theta$ is not divisible by p. All this shows that the critical representation (A, N) over L satisfies all the conditions of Lemma 3.3.4 if we take n as the bound for the classes of stable factors. Hence the number of generators for the group N is bounded by some $r = r(d, n, e)$. We emphasize that this r does not depend on the field L or even the ring R.

Thus the variety $\mathcal{X}$ over R is generated by representations of r-generated groups. Hence $r_b(\mathcal{X}) \leq r$, and since R is an arbitrary finitely generated subring of K, by Lemma 0.4.4 we also have $r_b(\mathcal{Y}) \leq r$. This completes the proof of the theorem. $\square$

3.5. Corollaries and related results

From Theorem 3.4.3 it follows, in particular, that the variety generated by a representation of a finite group is a variety of bounded rank. Keeping in mind the Oates–Powell theorem, one can naturally ask whether the variety generated by a representation of a finite group is actually a *Cross* variety. For ordinary representations the answer is positive in view of Theorem 3.1.2, but in general it is not the case. Indeed, let K be an infinite field of characteristic p. Consider the variety S^4 over K; then, by Proposition 1.1.7, S^4 is generated by its free representation $\mathrm{Fr}_4 S^4$. Hence S^4 is also generated by the corresponding faithful representation

$$\overline{\mathrm{Fr}_4 S^4} = (K F_4 / \Delta^4,\ F_4 / \mathrm{Ker}(\mathrm{Fr}_4 S^4)).$$

By 3.2.3 (i), $F_4 / \mathrm{Ker}(\mathrm{Fr}_4 S^4)$ is a nilpotent p-group, and hence it is finite. Thus S^4 is generated by a representation of a finite group. On the other hand, S^4 has infinitely many subvarieties [79] and therefore can not be Cross.

However, if the ground field is finite, the above question is answered in the affirmative.

3.5.1. Corollary. *The variety generated by a representation of a finite group over a finite field is Cross.*

P r o o f. The variety $\mathcal{X}$ generated by a representation of a finite group is locally finite, and by Theorem 3.4.3 it is finitely based. Therefore it suffices to show that if the ground field is finite, then $\mathcal{X}$ contains only a finite number of nonisomorphic critical representations. From the proof of Theorem 3.4.3 in prime characteristic, it follows that $\mathcal{X} \subseteq \mathcal{C}(d, s; e, m, c)$ for some d, s, e, m, c. Therefore Lemma 3.3.3 implies that the orders of critical representations from $\mathcal{X}$ are uniformly bounded in terms of s, d, m and char K. It remains to notice that every critical representation is cyclic, and

that a finite group has only a finite number of cyclic representations over a
given finite field. □

3.5.2. Corollary. *If a variety of group representations over a field of
prime characteristic is generated by a stable-by-finite representation, then it
is generated by a finite representation.*

Proof. Repeating the previous argument, we obtain that the orders
of critical representations in such a variety $\mathcal{X}$ are uniformly bounded. It
follows that $\mathcal{X}$ is generated by its free representation of some finite rank.
Since $\mathcal{X}$ is locally finite, this representation is finite. □

The following results show that certain properties of varieties, which in
general are quite distinct, turn out to be pairwise equivalent if we restrict
ourselves to locally finite or locally finite-dimensional varieties. Their proofs
are obtained by combining some arguments of Sections 2.2, 3.1, 3.3, 3.4 and
a few observations from [78].

3.5.3. Proposition. *Let $\mathcal{X}$ be a locally finite variety. Then the
following conditions are equivalent:*

(a) *$\mathcal{X}$ satisfies the maximum condition on subvarieties;*

(b) *$\mathcal{X}$ is generated by a finite-dimensional representation;*

(c) *$\mathcal{X}$ is generated by a finite representation;*

(d) *$\mathcal{X}$ is generated by a stable-by-finite representation;*

(e) *$\mathcal{X} \subseteq \mathcal{C}(d, s; e, m, c)$ for some d, s, e, m, c;*

(f) *$\mathcal{X}$ is a variety of bounded rank.*

Proof. (a) $\Longrightarrow$ (b). Let $\mathcal{X}$ satisfy the maximum condition on subvarieties. Consider the set of all subvarieties $\mathcal{X}_0$ of $\mathcal{X}$ such that $\mathcal{X}_0$ is generated
by a finite-dimensional representation. This set has a maximal element, say
$\mathcal{Y}$. Assume that $\mathcal{Y} \neq \mathcal{X}$. Then there exists a finitely generated representation $\rho \in \mathcal{X}$ not belonging to $\mathcal{Y}$. Since $\mathcal{X}$ is locally finite, ρ must be finite.
Therefore var$\{\rho, \mathcal{Y}\}$ can be generated by a finite-dimensional representa-
tion, which contradicts the maximality of $\mathcal{Y}$. Thus $\mathcal{Y} = \mathcal{X}$, whence $\mathcal{X}$ is
generated by a finite-dimensional representation.

(b) $\implies$ (c). Let $\mathcal{X}$ be generated by a finite-dimensional representation $\sigma = (W, H)$. Since $\mathcal{X}$ is locally finite, it is contained in some $\omega\Theta$, where Θ is a locally finite variety of groups. Denote $\exp\Theta = e$ and $\dim W = d$. By Lemma 2.4.2, $\mathcal{X}$ is generated by critical factors of σ. If $\rho = (V, G)$ is one of these factors, then $\dim V \leq d$. Choose in V a G-composition series

$$0 = V_0 \subset V_1 \subset \cdots \subset V_t = V \qquad (t \leq d) \tag{1}$$

and let $H_i = \mathrm{Ker}(V_{i+1}/V_i, G)$. Then G/H_i is an irreducible matrix group of degree $\leq d$ over K. Now the proof divides into several cases.

1) *K is an algebraically closed field of characteristic p.* Then G/H_i is an *absolutely irreducible* matrix group of degree $\leq d$. Since $\exp(G/H_i) \leq e$, the order $|G/H_i|$ of this group is bounded by some number depending only on d and e (see for instance [87, §23.3, Lemma 2]). The same is true for $|G/H|$ where $H = \cap_{i=1}^t H_i$.

As usual, let $O_p = O_p(G)$. Since $H \subseteq O_p$, we have $|G/O_p| \leq k$, where k depends only on d and e. Repeating literally the argument from the proof of Lemma 3.3.3, we obtain that the order of G is bounded by a number depending only on d, e, p and s, where s is the upper bound for the classes of stable factors of ρ. Since $s \leq \dim V \leq d$, we see that $|G|$ is bounded by some number $n = n(d, e, p)$.

Thus $\mathcal{X}$ is generated by representations of n-generated groups. Therefore $\mathcal{X}$ is generated by its free representation $\mathrm{Fr}_n\mathcal{X}$, which is finite.

2) *K is an arbitrary field of characteristic p.* Let $\bar{K}$ be the algebraic closure of K. Consider the map

$$\nu : \mathbb{M}(K) \to \mathbb{M}(\bar{K})$$

defined as in §0.4. A straightforward verification shows that $\mathcal{X}^\nu$ is a locally finite variety of group representations over $\bar{K}$ generated by the finite-dimensional representation $\sigma_{\bar{K}} = (\bar{K} \otimes_K W, H)$. By 1), $\mathcal{X}^\nu$ is generated by a finite representation τ. Since

$$\mathcal{X} = \mathcal{X}^{\nu\nu'} = \mathrm{var}_K(\mathcal{X}^\nu) = \mathrm{var}_K(\mathrm{var}_{\bar{K}}\tau) = \mathrm{var}_K\tau,$$

$\mathcal{X}$ is generated by a representation of a finite group, and so by a finite representation.

3) K *is a field of characteristic zero.* Since G/H_i is a matrix group over K of degree $\leq d$ and of exponent $\leq e$, by the Burnside theorem its order is bounded by a number depending only on d and e. The same is true for $|G/H|$, where $H = \cap H_i$. Now we note that H acts faithfully and stably in the series (1). Since $\mathrm{char}\,K = 0$, this implies that H must be a torsion-free nilpotent group. But G is finite, and so $H = \{1\}$. (This argument actually shows that *in a locally finite variety of group representations over a field of characteristic zero, every stable representation is trivial.*) Thus $|G| \leq n = n(d,e)$ and the proof can now be completed as in 1).

(d) $\implies$ (e). If $\mathrm{char}\,K = p$, this follows from Lemma 3.4.4. Let $\mathrm{char}\,K = 0$, and let $\mathcal{X}$ be a variety generated by a finite representation over K. By Corollary 3.1.5, $\mathcal{X}$ is contained in some class $\mathcal{C}(d; e, m, c)$, and it remains to note that every stable representation in $\mathcal{X}$ is trivial.

(e) $\implies$ (f). If $\mathrm{char}\,K = p$, this implication was established in the proof of Theorem 3.4.3. If $\mathrm{char}\,K = 0$, it is contained in Corollary 3.1.5.

Since the implications (c) $\implies$ (d) and (f) $\implies$ (a) are trivial, the proof of Proposition 3.5.3 is completed. $\square$

Let again $\mathcal{X}$ be a locally finite variety of group representations. We say that $\mathcal{X}$ is *ordinary* if $\mathcal{X} \subseteq \omega\Theta$, where Θ is a locally finite variety of groups with exponent not divisible by $\mathrm{char}\,K$. Combining the results of § 3.1 and Proposition 3.5.3, we obtain:

3.5.4. Proposition. *Let $\mathcal{X}$ be an ordinary locally finite variety of group representations. Then the following conditions are equivalent:*

(a)–(f) *from the preceding statement;*

(g) $\mathcal{X}$ *is a Cross variety;*

(h) $\mathcal{X}$ *contains only a finite number of simple representations;*

(k) $\mathcal{X}$ *has only a finite number of subvarieties;*

(l) $\mathcal{X}$ *satisfies the identity $v_d^{(n)}$ for some d and n;*

(m) $\mathcal{X} \subseteq \mathcal{C}(d; e, m, c)$ *where $\mathrm{char}\,K \nmid e$.* $\square$

Suppose now that $\mathcal{X}$ is *equal* to $\omega\Theta$ with Θ as above. Then the situation becomes completely transparent.

3.5.5. Proposition (Plotkin [78]). *Let $\mathcal{X} = \omega\Theta$ where Θ is a locally finite variety of groups whose exponent e is not divisible by char K. Then the conditions (a)-(m) are satisfied if and only if $\Theta = \mathcal{A}_e$.*

P r o o f. Let $\Theta = \mathcal{A}_e$, the variety of abelian groups of exponent e. Denote by $\bar{K}$ the algebraic closure of K and again consider the map ν : $\mathrm{M}(K) \to \mathrm{M}(\bar{K})$. Clearly $\mathcal{X}^\nu$ coincides with the variety $\omega\mathcal{A}_e$ over $\bar{K}$ and so is locally finite and ordinary. Hence $\mathcal{X}^\nu$ is generated by finite simple representations. Since an irreducible representation of an abelian group over an algebraically closed field is one-dimensional, it follows that $\mathcal{X}^\nu$ contains only a finite number of finite simple representations. Therefore the number of subvarieties of $\mathcal{X}^\nu$ is finite. Since the map ν is injective (Proposition 0.4.2), the number of subvarieties of $\mathcal{X}$ is finite as well. By 3.5.4, $\mathcal{X}$ is generated by a finite representation.

Conversely, let $\omega\Theta$ be generated by a finite representation. Then $\omega\Theta$ satisfies some multilinear identity $u(x_1,\ldots,x_n)$ (for example, the standard polynomial identity of degree $2d$ where d is the dimension of the generating representation). If $G \in \Theta$, then $\operatorname{Reg} G = (KG, G)$ satisfies $u(x_1,\ldots,x_n)$ and so $u(g_1,\ldots,g_n) = 0$ for any $g_i \in G$. Since $u(x_1,\ldots,x_n)$ is multilinear, it follows that $u(a_1,\ldots,a_n) = 0$ for all $a_i \in KG$. Thus for every $G \in \Theta$ the group algebra KG satisfies the identity $u(x_1,\ldots,x_n)$.

Now we prove that Θ must be abelian (cf. [80, pp.173-174]). Suppose the opposite. Then Θ contains a group A whose commutator subgroup A' is infinite. Let $G = A \times A \times \ldots$ be an infinite direct power of A. Since its group algebra KG satisfies some polynomial identity, by a theorem of Passman [73'], G has a normal subgroup of finite index whose commutator subgroup is finite. If H is such a subgroup, then $|G : H| < \infty$ implies that the natural projection of G into some direct factor A is surjective. But then A' is an epimorphic image of H', whence A' must be finite. This contradiction completes the proof. $\square$

Similar criteria can be proved for locally finite-dimensional varieties. If the ground field has prime characteristic, we can apply the previous propositions (since in prime characteristic every locally finite-dimensional variety is locally finite). Consider separately the case of characteristic zero.

3.5.6. Proposition. *Let $\mathcal{X}$ be a locally finite-dimensional variety of group representations over a field of characteristic zero. Then the following conditions are equivalent:*

    (a) *$\mathcal{X}$ satisfies the maximum condition on subvarieties;*

    (b) *$\mathcal{X}$ is generated by a finite-dimensional representation;*

    (c) *$\mathcal{X}$ is generated by a stable-by-finite representation;*

    (d) *$\mathcal{X}$ satisfies the identity $v_d^{(n)}$ for some d and n;*

    (e) *$\mathcal{X}$ is a variety of bounded rank.*

P r o o f. The implication (a) $\Longrightarrow$ (b) is proved as in Proposition 3.5.3. We show that (b) implies (c). Let $\mathcal{X}$ be generated by a finite-dimensional representation $\rho = (V, G)$. Since $\mathcal{X} \subseteq S^n \times \Theta$ for some n and some locally finite variety of groups Θ, G possesses a normal subgroup N such that $(V, N) \in S^n$ and $G/N \in \Theta$. Let

$$0 = V_0 \subset V_1 \subset \cdots \subset V_t = V$$

be the upper N-stable series in V, then each V_i is invariant under G. Denote by H_i the kernel of the naturally arising representation $(V_{i+1}/V_i, G)$. The group G/H_i can be regarded as a matrix group over the ground field K. Since $N \subseteq H_i$, we have $G/H_i \in \Theta$ and hence $\exp(G/H_i) < \infty$. By the Burnside theorem, G/H_i is finite. Denote $H = \cap H_i$. Then G/H is finite and, on the other hand, H acts stably in (2). Thus ρ is stable-by-finite.

By Lemma 3.2.2, (c) implies (d). By Theorem 3.4.3, (e) implies (a). So it remains to show that (e) follows from (d). Let $v_d^{(n)}$ be satisfied in $\mathcal{X}$. We again use the fact that $\mathcal{X} \subseteq S^n \times \Theta$ with a locally finite Θ. Denote by $\mathcal{V}$ the variety of all representations from $S^n \times \Theta$ satisfying the identity $v_d^{(n)}$. Repeating literally the proof of Theorem 3.4.3 in the case of characteristic zero, we obtain that $\mathcal{V}$ is a variety of bounded rank. It remains to notice that $\mathcal{X} \subseteq \mathcal{V}$. $\square$

There are many open problems closely related to the considerations of the present chapter. Here we mention only two of them.

3.5.7. Problem. *Does every finite-dimensional representation have a finite basis of identities?*

Undoubtedly, this problem is one of the most important and intriguing questions of the theory of varieties of group representations.

3.5.8. Problem. *Does every representation of a finite group over a noetherian ring K have a finite basis of identities? In particular, is it true if K is the ring of integers? if the module of the representation is finite?*

We believe that the methods used in proving Theorem 3.4.3, combined with more sophisticated ring-theoretic machinery, may give an approach to the latter problem.

Chapter 4

FURTHER TOPICS

Our final chapter deals with a variety of themes at varying levels of detail. It provides a selection of results which, taken together, illustrate the diversity of the field and the broad range of techniques used.

In § 4.1–4.2 we continue to investigate the finite basis problem for identities of group representations. More specifically, we are concerned with questions of the following type: for a given *multilinear* identity, is a variety satisfying this identity finitely based?

It is well known that every identity of a linear algebra over a field of characteristic zero is equivalent to a system of multilinear identities, which can be effectively derived from the initial one. For group representations the corresponding statement is not true (otherwise, by Theorem 1.2.4, every variety would be homogeneous) and, in general, multilinear identities do not play here such a prevalent role. But if a variety does happen to be determined by multilinear identities, or at least satisfies some identity of this sort, then one can immediately derive significant consequences. For example, in many cases such a variety is finitely based.

The main result of § 4.1 states that in certain respects the behavior of multilinear identities of group representations is closely related to the behavior of those of associative algebras. In particular, from this fact and a recent outstanding result of Kemer [40], we deduce that every system of multilinear identities of group representations over a field of characteristic zero is finitely based. In § 4.2 we prove a rather old theorem of Cohen [10] which implies that every representation of an abelian group over a noetherian ring is finitely based. A far-going generalization of this theorem has been recently announced by Krasil'nikov [47]. The exact statement of

his result and some related questions and facts are discussed at the end of
the section.

Sections 4.3–4.4 deal with pure varieties. Recall that a variety $\mathcal{X}$ of
group representations over an integral domain R is called *pure* if $RF/\mathrm{Id}\,\mathcal{X}$
is a torsion-free R-module. The interest in pure varieties is motivated by the
fact that they are in a natural one-to-one correspondence with all varieties
over the field of quotients of R. Section 4.3 is mainly devoted to the proof
of an unpublished result of G. M. Bergman on products of ideals in RF
which implies that if R is a Dedekind domain, then the product of two pure
varieties over R is also a pure variety (this fact was earlier proved by the
author [94, Corollary 9.9]). On the other hand, at the end of the section we
sketch an example showing that over an arbitrary domain this, in general,
is not true. In § 4.4 we study one concrete series of varieties and prove that
the intersection of pure varieties need not be pure even over $\mathbb{Z}$.

Finally, § 4.5 provides a brief overview of some applications of our theory to varieties of groups, varieties of rings, and dimension subgroups.

The exposition in this chapter (especially in § 4.5) is not as complete
as in the previous ones: several proofs are omitted or merely outlined. The
ground ring is, in general, arbitrary.

4.1. Multilinear identities and partial linearizations

What is a multilinear identity of group representations?

On the one hand, it is natural to say that an element $u(x_1,\ldots,x_n) \in$
KF is multilinear if

$$u(x_1,\ldots,x_n) = \sum_{\sigma \in S_n} \lambda_\sigma x_{\sigma(1)} \ldots x_{\sigma(n)} \qquad (1)$$

where $\lambda_\sigma \in K$. As an example, one can mention the so-called standard
polynomial of degree n, that is, the polynomial

$$s_n(x_1,\ldots,x_n) = \sum_{\sigma \in S_n} (\mathrm{sign}\,\sigma) x_{\sigma(1)} \ldots x_{\sigma(n)}.$$

According to the Amitsur–Levitzky Theorem, it is an identity of any representation over a field whose dimension does not exceed $n/2$.

On the other hand, in Chapter 1 we adopted another definition: an element $u(x_1, \ldots, x_n) \in KF$ was said to be multilinear if

$$u(x_1, \ldots, x_n) = \sum_{\sigma \in S_n} \lambda_\sigma z_{\sigma(1)} \cdots z_{\sigma(n)} \qquad (2)$$

where $\lambda_\sigma \in K$ and $z_i = x_i - 1$. In Chapter 1 this approach was completely justified, for it was closely related to the idea of homogeneity and led to direct connections with varieties of associative algebras. Now we will show that from the standpoint of the finite basis problem, the "z_i-approach" is also more natural and reasonable than "x_i-approach".

Let $u = u(x_1, \ldots, x_n) \in KF$ be a *polynomial*, that is, u does not involve negative powers of the x_i. We say that u is *quasi-multilinear*, if the degree of u in each x_i does not exceed 1. For example, every element of the form (1) or (2) is quasi-multilinear. More generally, every element of the form

$$\sum_{\sigma \in S_n} \lambda_\sigma t_{\sigma(1)} \cdots t_{\sigma(n)} \qquad (3)$$

where $\lambda_\sigma \in K$ and $t_i = \alpha_i x_i + \beta_i$ $(\alpha_i, \beta_i \in K, \ i = 1, \ldots, n)$, is a quasi-multilinear polynomial.

4.1.1. Lemma (Krasil'nikov). *Every quasi-multilinear polynomial is equivalent, as an identity of group representations, to a finite set consisting of polynomials multilinear in z_i and a scalar from K.*

P r o o f. Let $u = u(x_1, \ldots, x_n) \in KF$ be a quasi-multilinear polinomial. Denote $u(1, x_2, \ldots, x_n) = v(x_2, \ldots, x_n) = v$, and let $w = u - v$. Both the polynomials v and w are quasi-multilinear and, moreover, w is homogeneous of degree 1 in $z_1 = x_1 - 1$. Since v is a consequence of u, we see that w is a consequence of u as well. Therefore $u = v + w$ is equivalent to the set $\{v, w\}$. The same argument may be applied to v and w, and so on.

To make the last statement more precise, consider the set S of elements $f \in KF$ satisfying the following three conditions:

(a) f can be written as a linear combination of terms $t_1 t_2 \ldots t_m$ (plus possibly a constant term), where each factor t_i is equal to either some x_k or some z_k;

(b) every term $t_1 t_2 \ldots t_m$ is a polynomial of degree ≤ 1 in each x_k;

(c) if z_k is involved in some term $t_1 t_2 \ldots t_m$, then it is involved in all the other terms (that is, f is homogeneous of degree 1 in each z_k it involves).

Now we note that our quasi-multilinear polynomial $u = u(x_1, \ldots, x_n)$ certainly belongs to S. Present u as a sum of terms $t_1 t_2 \ldots t_m$ satisfying (a)–(c). If each t_i in every term is equal to some z_k then, by (c), u is multilinear in the variables z_k and there is nothing to prove. Otherwise one can find t_i which is equal to some x_k, so that the corresponding term does not involve z_k. Applying to u the above procedure, we get

$$u = v + w \quad \text{where} \quad v = u(x_1, \ldots, x_{k-1}, 1, x_{k+1}, \ldots, x_n).$$

Both polynomials v and w belong to S, the number of variables in v is $n-1$, and w is now *homogeneous of degree 1 in z_k*. Using the obvious induction, we eventually obtain a finite set of polynomials multilinear in the z_k, and possibly, the scalar $u(1, \ldots, 1)$. $\square$

It is easy to understand that if $\mathcal{X}$ is a variety of group representations over a ring K and $\lambda \in K$ is an "identity" of this variety, then the study of $\mathcal{X}$ can be reduced to the study of the corresponding variety over the factor-ring $K/(\lambda)$ (at least when the finite basis problem is concerned). Together with Lemma 4.1.1, this shows that every identity of the form (3) can be reduced to identities of the form (2) and explains why *by a multilinear identity of group representations we always mean an identity multilinear in the variables $z_i = x_i - 1$*.

As usual, let $K\langle Z \rangle$ be the subalgebra (without 1) of KF generated by the set $Z = \{z_1, z_2, \ldots\}$. It is an absolutely free associative algebra and Z is a free generating set for $K\langle Z \rangle$. *We emphasize that $K\langle Z \rangle$ does not contain any nonzero scalars.* Every element of $K\langle Z \rangle$ can be uniquely written as a polynomial

$$f = f(z_1, \ldots, z_n),$$

with constant term 0, in the noncommuting variables z_i. From now on, by a polynomial we will always mean an element from $K\langle Z \rangle$; in particular, the elements of the form $\lambda z_{i_1} \ldots z_{i_n}$ (where $\lambda \in K$ and $n > 0$) will be called *monomials.* Recall that a polynomial $f(z_1, \ldots, z_n)$ is *multihomogeneous* if, for each z_i, its monomials all have the same degree in z_i. Thus a polynomial is multilinear if and only if it is multihomogeneous of degree 1 in each of the z_i which are involved in it.

Let $M \subseteq K\langle Z \rangle$ be an arbitrary set of polynomials and let $T(M)$ be the T-ideal of $K\langle Z \rangle$ generated by M. Of course, in general $T(M) \not\subseteq \mathrm{Id}(M)$. There arises a question: does there exists a "natural" extension M' of M such that $T(M) \subseteq \mathrm{Id}(M')$? In other words, how is the deducibility of identities in the "ring-theoretic" sense related to that in the "group-representation" sense? We will investigate this question via the notion of partial linearization of identities.

Following [106], for every $i = 1, 2, \ldots$, $k = 0, 1, 2, \ldots$ and $z \in Z = \{z_1, z_2, \ldots\}$, let $\Delta_i^k(z)$ denote the K-linear transformation of $K\langle Z \rangle$ defined as follows. Let c be any monomial from $K\langle Z \rangle$ and let $m = \deg_{z_i} c$. Then

(a) if $k = 0$, then $c\Delta_i^k(z) = c$;

(b) if $k > m$, then $c\Delta_i^k(z) = 0$;

(c) if $1 \leq k \leq m$, then $c\Delta_i^k(z) = c_1 + c_2 + \cdots + c_{\binom{m}{k}}$, where the summands are all the possible monomials obtainable from c by substituting z instead of k occurrences of z_i.

Example. Let $c = z_1 z_2^2 z_3 z_2$. Then

$$c\Delta_2^1(z) = z_1 z_2^2 z_3 z + z_1 z_2 z z_3 z_2 + z_1 z z_2 z_3 z_2,$$

$$c\Delta_2^2(z) = z_1 z_2 z z_3 z + z_1 z z_2 z_3 z + z_1 z^2 z_3 z_2,$$

$$c\Delta_2^3(z) = z_1 z^2 z_3 z,$$

$$c\Delta_2^4(z) = c\Delta_2^5(z) = \cdots = 0.$$

The maps $\Delta_i^k(z)$ are called the *operators of partial linearization*. For arbitrary $\Delta_i^k(z)$ and arbitrary $f \in \mathcal{F}$, we say that $\Delta_i^k(z)$ acts *properly* on f if the variable z is not involved in f. Furthermore, let f be a multihomogeneous polynomial and $m = \deg_{z_i} f$. We say that $\Delta_i^k(z)$ acts *strictly* on f if $0 < k < m$.

For every $M \subseteq K\langle Z \rangle$ let $M\Delta$ be the smallest K-submodule of $K\langle Z \rangle$ containing M and closed under all $\Delta_i^k(z)$. The following theorem gives one possible answer to the raised question.

4.1.2. Theorem (Vovsi and Volkov [102]). *If $M \subseteq K\langle Z \rangle$, then*

$$T(M) \subseteq \mathrm{Id}(M\Delta).$$

To prove this theorem, we need two elementary lemmas concerning the operators $\Delta_i^k(z)$.

4.1.3. Lemma. *If $f \in M$, then all multihomogeneous components of f belong to $M\Delta$.*

P r o o f. Let $f = f(z_1, \ldots, z_n) = f_1 + f_2$, where f_1 is homogeneous in z_1 of degree m and $\deg_{z_1} f_2 < m$. Then

$$f\Delta_1^m(z_{n+1}) = f_1\Delta_1^m(z_{n+1}) + f_2\Delta_1^m(z_{n+1})$$
$$= f_1(z_{n+1}, z_2, \ldots, z_n),$$

while $f_1(z_{n+1}, z_2, \ldots, z_n)\Delta_{n+1}^m(z_1) = f_1$ and $f_2 = f - f_1$. Hence $f_1 \in M\Delta$ and $f_2 \in M\Delta$. It is clear that continuing this process we eventually come to the multihomogeneous components of f. $\square$

4.1.4. Lemma. *Let $f = f(z_1, \ldots, z_n)$ be a multihomogeneous polynomial from M and let $c_i^j \in KF$ ($i = 1, \ldots, n$; $j = 1, \ldots, s_i$). Then*

$$f(c_1^1 + \cdots + c_1^{s_1}, \ldots, c_n^1 + \cdots + c_n^{s_n}) = \sum_r f_r(c_{i_1(r)}^{j_1(r)}, \ldots, c_{i_t(r)}^{j_t(r)})$$

where all the f_r are multihomogeneous polynomials from $M\Delta$.

P r o o f. It is well known that if $\deg_{z_i} f = m$, then

$$f(z_1, \ldots, z_i + z, \ldots, z_n) = \sum_{k=0}^{m} f\Delta_i^k(z)$$

(see [106, p.20]), and since f is multihomogeneous, all the $f\Delta_i^k(z)$ are multihomogeneous as well. Applying this remark several times, we eventually obtain that for arbitrary pairwise distinct variables $z_i^j \in Z$ ($i = 1, \ldots, n$; $j = 1, \ldots, s_i$)

$$f(z_1^1 + \cdots + z_1^{s_1}, \ldots, z_n^1 + \cdots + z_n^{s_n}) = \sum_r f_r(z_{i_1(r)}^{j_1(r)}, \ldots, z_{i_t(r)}^{j_t(r)})$$

where each f_r is a multihomogeneous polynomial obtained from f by applying a finite number of operators $\Delta_i^k(z)$. Hence $f_r \in M\Delta$ and it remains to replace the z_i^j by the corresponding c_i^j (a relation between free generators of $K\langle Z \rangle$ is an identical relation of every associative algebra). $\square$

Note. Suppose that all the z_i, z_i^j in the proof of Lemma 4.1.4 are pairwise distinct. Then it is clear that none of the $\Delta_i^k(z)$ in this proof is applied to a polynomial already involving z. In other words, we can ensure that each of the f_r is obtained by finitely many applications of operators $\Delta_i^k(z)$, each operator acting *properly* in each of these applications.

P r o o f   o f   T h e o r e m   4.1.2. 1) Let M' be the set of all multihomogeneous components of polynomials from M. By Lemma 4.1.3, $M' \subseteq M\Delta$, and since $T(M) \subseteq T(M')$, it suffices to prove that $T(M') \subseteq \mathrm{Id}(M'\Delta)$. In other words, we may (and will) assume M to consist of multihomogeneous polynomials.

2) Every element of $T(M)$ is a sum of elements of the form $af^\psi b$, where $a, b \in \mathcal{F}$, $f \in M$, $\psi \in \mathrm{End}(K\langle Z \rangle)$. Hence it is enough to show that $f^\psi \in \mathrm{Id}(M\Delta)$. If $f = f(z_1, \ldots, z_n)$, then $f^\psi = f(z_1^\psi, \ldots, z_n^\psi)$. The polynomials z_i^ψ can be written as sums of monomials

$$z_i^\psi = c_i^1 + c_i^2 + \cdots + c_i^{s_i} \qquad (i = 1, \ldots, n).$$

By Lemma 4.1.4,

$$f(z_1^\psi, \ldots, z_n^\psi) = \sum_r f_r(c_{i_1(r)}^{j_1(r)}, \ldots, c_{i_t(r)}^{j_t(r)})$$

where all the $f_r(z_1, \ldots, z_t)$ are multihomogeneous and belong to $M\Delta$. Thus it suffices to show that *if $f(z_1, \ldots, z_n)$ is a multihomogeneous polynomial from $M\Delta$, then $f(c_1, \ldots, c_n) \in M\Delta$ for any monomials $c_1, \ldots, c_n$.*

3) Let $m_i = \deg_{z_i} f$. Arrange the numbers m_i in the nonincreasing order: $m_{i_1} \geq m_{i_2} \geq \cdots \geq m_{i_n}$. Then the sequence

$$t(f) = (m_{i_1}, m_{i_2}, \ldots, m_{i_n}, 0, 0, \ldots)$$

is called the *type* of f. Evidently the set of all types is well ordered under the lexicographic ordering. Furthermore, it is easy to see that if $\Delta_i^k(z)$ acts properly and strictly on f (that is, if z is not involved in f and $0 < k < m_i$), then $t(f\Delta_i^k(z)) < t(f)$ (because the component m_i of $t(f)$ splits into two nonzero components k and $m_i - k$).

Suppose that our assertion is false. Then there exists a multihomogeneous polynomial $f(z_1, \ldots, z_n) \in M\Delta$ of the smallest type such that

$$f(c_1, \ldots, c_n) \notin \mathrm{Id}(M\Delta) \qquad (4)$$

for some monomials $c_1, \ldots, c_n$. We will assume that the set of monomials $c_1, \ldots, c_n$ satisfying (4) is minimal with respect to the number $\sum_{i=1}^n \deg c_i$. Thus our choice of the polynomial f and the monomials $c_1, \ldots, c_n$ guarantees the following:

(i) if $g(z_1, \ldots, z_k) \in M\Delta$ is a multihomogeneous polynomial and $t(g) < t(f)$, then $g(d_1, \ldots, d_k) \in \mathrm{Id}(M\Delta)$ for any monomials $d_1, \ldots, d_k$;

(ii) if $d_1, \ldots, d_n$ are monomials satisfying the condition $\sum \deg d_i < \sum \deg c_i$, then $f(d_1, \ldots, d_n) \in \mathrm{Id}(M\Delta)$.

Every c_i can be written as $\lambda_i z_{i_1} z_{i_2} \ldots z_{i_r}$, where $\lambda_i \in K$. Denoting $\bar{c}_i = z_{i_1} z_{i_2} \ldots z_{i_r}$, we have

$$c_i = \lambda_i \bar{c}_i \quad (i = 1, \ldots, n).$$

Define $\varphi \in \mathrm{End}\, F$ by the rule

$$x_i^\varphi = \begin{cases} x_{i_1} x_{i_2} \ldots x_{i_r} & \text{if } i = 1, \ldots, n; \\ x_i & \text{if } i > n. \end{cases}$$

Then

$$\begin{aligned} z_i^\varphi &= x_{i_1} x_{i_2} \ldots x_{i_n} - 1 \\ &= (z_{i_1} - 1)(z_{i_2} - 1) \ldots (z_{i_r} - 1) - 1 \\ &= \bar{c}_i + w \end{aligned}$$

where w is a sum of monomials of degree at most $r - 1$.

Since $f \in M\Delta$, it follows that $f^\varphi \in \mathrm{Id}(M\Delta)$. On the other hand, by Lemma 4.1.4 we have

$$f^\varphi = f(z_1^\varphi, \ldots, z_n^\varphi) = f(\bar{c}_1, \ldots, \bar{c}_n) + \Sigma_1 + \Sigma_2. \qquad (5)$$

Here Σ_1 is a sum of elements of the form $f(d_1, \ldots, d_n)$ where the d_i are monomials satisfying the condition $\sum \deg d_i < \sum \deg c_i$, and Σ_2 is a sum of elements of the form $g(d_1, \ldots, d_s)$, each $g(z_1, \ldots, z_s)$ being a multihomogeneous polynomial obtained from f by finitely many applications of operators $\Delta_i^k(z)$ with each application involving proper action (see the note after Lemma 4.1.4) and at least one application being strict. (We emphasize that $f(\bar{c}_1, \ldots, \bar{c}_n)$ and the summands of Σ_1 are also obtained by means of operators $\Delta_i^k(z)$, but in those cases no strict action is involved).

Every summand from Σ_1 belongs to $\mathrm{Id}(M\Delta)$ by (ii). Every summand from Σ_2 belongs to $\mathrm{Id}(M\Delta)$ by (i), for if $\Delta_i^k(z)$ acts properly and strictly on f, then $t(f\Delta_i^k(z)) < t(f)$. Since $f^\varphi \in \mathrm{Id}(M\Delta)$, it follows from (5) that $f(\bar{c}_1, \ldots, \bar{c}_n) \in \mathrm{Id}(M\Delta)$. Hence

$$f(c_1, \ldots, c_n) = f(\bar{c}_1, \ldots, \bar{c}_n) \prod_{i=1}^{n} \lambda_i^{m_i} \in \mathrm{Id}(M\Delta)$$

which is impossible in view of (4). $\square$

4.1.5. Corollary. *If $M \subseteq K\langle Z \rangle$ is such that $M\Delta \subseteq \mathrm{Id}(M)$ (for example, if M consists of multilinear elements), then $T(M) \subseteq \mathrm{Id}(M)$.* $\square$

The above results make it possible to apply the theory of varieties of associative algebras to the solution of several questions on multilinear identities of group representations. One such application can be demonstrated right now, while the others will be presented in the next section.

4.1.6. Corollary. *Let $M \subseteq K\langle Z \rangle$ be a set of multilinear polynomials. If $T(M)$ is finitely generated as a T-ideal, then $\mathrm{Id}(M)$ is finitely generated as a completely invariant ideal.*

Proof. By hypothesis, $T(M) = T(M_0)$ for some finite subset M_0 of M. It follows from Corollary 4.1.5 that $M \subseteq T(M) = T(M_0) \subseteq \mathrm{Id}(M_0)$, whence $\mathrm{Id}(M) = \mathrm{Id}(M_0)$. $\square$

By Kemer's theorem [40], every set of multilinear identities of associative algebras over a field of characteristic zero is finitely based. Together with Corollary 4.1.6, this yields the following (*a priori* far from evident) important consequence.

4.1.7. Corollary. *Every set of multilinear identities of group representations over a field of characteristic zero is finitely based.* $\square$

Since examples of non-finitely-based varieties of group representations over any K are commonly known (see 0.3.3), it follows that even over a field of characteristic zero there exist varieties which cannot be determined by multilinear identities.

4.1.8. Problem. *Does every multilinear identity of group representations determine a Specht variety?*

As usual with Specht varieties, this problem makes sense only if the ground ring is noetherian. It seems unlikely that the answer is positive (even over a field of zero characteristic), but we do not have any definite arguments. Some particular cases of the problem will be solved in the next section.

4.2. Identities of representations of abelian groups and related results

Consider one of the simplest multilinear identities

$$z_1 z_2 - z_2 z_1. \tag{1}$$

Since $z_1 z_2 - z_2 z_1 = x_1 x_2 - x_2 x_1$, it is clear that a representation $\rho = (V, G)$ satisfies (1) if and only if the group $G/\mathrm{Ker}\,\rho$ is abelian. Hence (1) determines the variety $\omega\mathcal{A}$, where $\mathcal{A}$ is the variety of abelian groups.

4.2.1. Theorem. *Over an arbitrary noetherian ring K, the variety of group representations defined by the identity (1) is Specht. In other words, every representation of an abelian group has a finite basis of identities.*

This theorem is a consequence of a ring-theoretic result of Cohen [10] whose proof incorporates some ideas of Higman [34] on closure operations and partially well-ordered sets. We begin by proving this result.

An *algebraic closure operation* on a set C is a map assigning to each subset X of C a subset $\overline{X}$ of C such that:

(i) $X \subseteq \overline{X} \subseteq \overline{Y} \subseteq \overline{\overline{Y}}$ whenever $X \subseteq Y$;

(ii) if $x \in \overline{X}$, then $x \in \overline{X_0}$ for some finite $X_0 \subseteq X$.

For example, $\overline{X}$ may be the subalgebra generated by a subset X of some algebra, the ideal generated by a subset X of a ring, etc. A subset $X \subseteq C$ is called *closed* if $X = \overline{X}$. An algebraic closure operation has *finite basis property* (f.b.p.) if every closed set is the closure of a finite subset.

A partially ordered set $(P, \leq) = P$ is said to be *partially well-ordered* if for every sequence of its elements $p_1, p_2, \ldots$ there exist numbers $i < j$ such that $p_i \leq p_j$. This notion was introduced by Erdős and Rado, and it was Higman who realized its connection with the finite basis property. Namely, for a partially ordered set P one can define the *natural closure operation on P* by letting

$$\overline{X} = \{p \,|\, \exists x \in X : x \leq p\}$$

for any subset X of P. Then a straightforward verification shows:

4.2.2. Lemma (Higman [34]). *The following conditions on a partially ordered set P are equivalent:*

(i) *P is partially well-ordered;*

(ii) *every infinite sequence of elements of P has an infinite nondecreasing subsequence;*

(iii) *there exists neither an infinite strictly decreasing sequence in P, nor an infinity of mutually incomparable elements of P;*

(iv) *the natural closure operation on P has f.b.p.;*

(v) *the closed subsets of P satisfy the ascending chain condition.* □

Note. Since the property of a set to be partially well-ordered turns out to be virtually equivalent to the f.b.p., in the subsequent arguments one could easily avoid mentioning the second property explicitly (see e.g. [3, § 5.2]). We prefer to follow the original Cohen's approach and use both these notions because, to our taste, this makes the exposition more transparent.

4.2.3. Example. Let $\mathbf{A}$ be the set of all finite sequences of non-negative integers, including the empty sequence. On this $\mathbf{A}$ we define two orderings:

$(i_1, \ldots, i_m) < (j_1, \ldots, j_n)$ if $m < n$, or $m = n$ and there exists r such that $i_r < j_r$ but $i_s = j_s$ for all $s > r$ (i.e. sequences of the same length are ordered lexicographically on the right);

$(i_1, \ldots, i_m) \preceq (j_1, \ldots, j_n)$ if $m \leq n$ and there exists a one-to-one order preserving map $\varphi : \{1, \ldots, m\} \to \{1, \ldots, n\}$ such that $i_r \leq j_{\varphi(r)}$ for all r (in other words, the first sequence is majorized by some subsequence of the second one).

It is easy to see that $(\mathbf{A}, \leq)$ is well-ordered, $(\mathbf{A}, \preceq)$ is partially ordered and the identity map is an order preserving map from $(\mathbf{A}, \preceq)$ to $(\mathbf{A}, \leq)$. Furthermore, by Theorem 4.3 from [34], $(\mathbf{A}, \preceq)$ is partially well-ordered.

Let C be a set with an algebraic closure operation and P any partially ordered set. For any subset $X \subseteq C \times P$ define its closure $\overline{X}$ by the rule: $(c, p) \in \overline{X}$ if and only if there exist $(c_1, p_1), \ldots, (c_n, p_n) \in X$ such that

$$c \in \overline{\{c_1, \ldots, c_n\}} \quad \text{and} \quad p_i \leq p \quad (1 \leq i \leq n). \tag{2}$$

It is easy to see that the map $X \mapsto \overline{X}$ is an algebraic closure operation on $C \times P$. We say that it is *induced* by the closure operation on C and the partial ordering on P.

4.2.4. Lemma. *If the closure operation on C has f.b.p. and P is partially well-ordered, then the induced operation on $C \times P$ has f.b.p.*

Proof. Let X be a closed subset of $C \times P$. For any $p \in P$ define $X(p) \subseteq C$ by the rule

$$X(p) = \{c \mid (c, p) \in X\}.$$

By the definition of the induced closure operation, $X(p)$ is a closed subset of C, and if $p < q$ then $X(p) \subseteq X(q)$. We order the collection of sets $\{X(p) \mid p \in P\}$ by letting

$$X(p) < X(q) \iff p < q \quad \text{and} \quad X(p) = X(q) \quad \text{as sets.}$$

Then for any p there exists q with $X(p) \geq X(q)$ and $X(q)$ minimal. Indeed, otherwise there would be an infinite sequence $X(p_1) > X(p_2) > \ldots$, and so an infinite strictly decreasing sequence $p_1 > p_2 > \ldots$, which is impossible because P is partially well-ordered.

We show now that there are only finitely many minimal elements in $\{X(p) \mid p \in P\}$. Assume the contrary, and let $X(p_1), X(p_2), \ldots$ be an infinite sequence of pairwise distinct minimal elements. Since P is partially well ordered, from the sequence $p_1, p_2, \ldots$ one can choose an increasing subsequence. Assume $p_1 \leq p_2 \leq \ldots$ is already such a sequence. Since the sets $X(p_i)$ are pairwise distinct, so are the p_i. Hence

(i) $X(p_i) \subseteq X(p_{i+1})$ because $p_i < p_{i+1}$;

(ii) $X(p_i) \neq X(p_{i+1})$ because $X(p_{i+1})$ is minimal.

Thus we have found an infinite sequence

$$X(p_1) \subset X(p_2) \subset \cdots \subset X(p_n) \subset \ldots$$

of closed subsets of C where all inclusions are strict. This is impossible in view of f.b.p. in C.

Let $X(p_1), \ldots, X(p_r)$ be all the minimal elements of $\{X(p) \mid p \in P\}$. Since C has f.b.p., $X(p_i) = \overline{Y_i}$ for some finite set Y_i. Now take an arbitrary $(c, p) \in X$. Then $c \in X(p)$ and, for some i, $X(p_i) \leq X(p)$. Therefore $c \in X(p_i) = \overline{Y_i}$. From the definition of the closure operation on $C \times P$, it follows that $(c, p) \in \overline{Y_i \times p_i}$. Hence X is the closure of the *finite* set $\bigcup_{i=1}^{n}(Y_i \times p_i)$, as required. $\square$

As usual, let $K[X] = K[x_1, x_2, \ldots]$ be the ring of polynomials over K in a countable number of (commuting!) variables x_i. Denote by Φ the set of all order preserving injection from the set $\mathbb{N}$ of natural numbers to itself. An ideal I of $K[X]$ is called a Φ-*ideal* if

$$f(x_1, \ldots, x_n) \in I \implies f(x_{\varphi(1)}, \ldots, x_{\varphi(n)}) \in I$$

whenever $f \in K[X]$ and $\varphi \in \Phi$. We can now prove the key result of this section which is a variation on the theme of the Hilbert Basis Theorem.

4.2.5. Theorem (Cohen [10]). *If K is noetherian, then $K[X]$ satisfies the ascending chain condition on Φ-ideals.*

Proof. 1) Let $\mathbf{A}$ be the set defined in 4.2.3. We define the *weight* of a monomial $\lambda x_1^{i_1} \ldots x_n^{i_n}$ to be the sequence $(i_1, \ldots, i_n) \in \mathbf{A}$ (for example, the weight of the monomial $x_3^2 x_4 x_6^3$ is $(0, 0, 2, 1, 0, 3)$). The *leading term* of a polynomial $f \in K[X]$ is its monomial of maximal weight in $(\mathbf{A}, \leq)$, and the weight wt f of f is the weight of its leading term. Define a map $\theta : K[X] \to K \times \mathbf{A}$ by letting

$$f^\theta = (\lambda, \text{ wt } f)$$

where λ is the leading coefficient of f. The closure operation on K, in which the closed sets are ideals, has f.b.p. because K is noetherian. By Lemma 4.2.4, the closure operation on $K \times \mathbf{A}$ induced from the closure operation on K and the ordering $\preceq$ on $\mathbf{A}$ has f.b.p.

2) We show that if I is a Φ-ideal of R, then I^θ is a closed subset of $K \times \mathbf{A}$. Let (λ, a), where $\lambda \in K$ and $a \in \mathbf{A}$, belong to the closure of I^θ. Then, according to (2), there exist $(\lambda_1, a_1), \ldots, (\lambda_n, a_n) \in I^\theta$ such that

$$\lambda \in K\lambda_1 + \cdots + K\lambda_n \quad \text{and} \quad a_k \preceq a \quad (1 \leq k \leq n). \tag{3}$$

Let $a = (i_1, \ldots, i_m)$ and $a_k = (i_1(k), \ldots, i_{m_k}(k))$, $1 \leq k \leq n$. Since $(\lambda_k, a_k) \in I^\theta$, there exist polynomials $f_1, \ldots, f_n \in I$ such that $f_k^\theta = (\lambda_k, a_k)$, that is, the leading term of f_k is

$$\lambda_k x_1^{i_1(k)} \ldots x_{m_k}^{i_{m_k}(k)} \tag{4}$$

for each $k = 1, \ldots, n$. One has to find a polynomial $g \in I$ whose leading term is $\lambda x_1^{i_1} \ldots x_m^{i_m}$.

By (3), $\lambda = \alpha_1 \lambda_1 + \cdots + \alpha_n \lambda_n$ for some $\alpha_1, \ldots, \alpha_n \in K$. Fix an arbitrary k between 1 and n and note that since $\alpha_k \preceq a$, there exists an order preserving injection $\varphi : \{1, \ldots, m_k\} \to \{1, \ldots, m\}$ such that

$$i_r(k) \leq i_{\varphi(r)} \quad \text{for} \quad r = 1, \ldots, m_k. \tag{5}$$

We can extend φ to an order preserving injection from $\mathbb{N}$ to itself; the resulting extension again denote by φ. The polynomial $g_k = f_k(x_{\varphi(1)}, x_{\varphi(2)}, \ldots)$ belongs to I, and since (4) is the leading term of f_k, it is clear that

$$\lambda_k x_{\varphi(1)}^{i_1(k)} \ldots x_{\varphi(m_k)}^{i_{m_k}(k)} \tag{6}$$

is the leading term of g_k. By (5), one can find a monomial c_k in the variables $x_1, \ldots, x_m$ such that the leading term of the polynomial $\alpha_k g_k c_k$ is equal to $\alpha_k \lambda_k x_1^{i_1} \ldots x_m^{i_m}$. This is true for each $k = 1, \ldots, n$, hence the leading term of the polynomial $g = \alpha_1 g_1 c_1 + \cdots + \alpha_n g_n c_n$ is

$$(\alpha_1 \lambda_1 + \cdots + \alpha_n \lambda_n) x_1^{i_1} \ldots x_m^{i_m} = \lambda x_1^{i_1} \ldots x_m^{i_m},$$

while g together with $g_1, \ldots, g_n$ belongs to I, as required.

3) Thus I^θ is a closed subset of $K \times \mathbf{A}$ and $K \times \mathbf{A}$ has f.b.p. Therefore $I^\theta = \overline{\{f_1^\theta, \ldots, f_n^\theta\}}$ for some polynomials $f_1, \ldots, f_n \in I$. Denote by I_0 the Φ-ideal generated by the f_i. The arguments from 2) show that for any $f \in I$ there exists a polynomial $g \in I_0$ such that $f^\theta = g^\theta$, that is, the leading terms of f and g coincide. But then $\mathrm{wt}(f - g) < \mathrm{wt}\, f$, and since $(\mathbf{A}, \leq)$ is well ordered, it follows by induction that $f \in I_0$. Thus $I = I_0$ and so every Φ-ideal of $K[X]$ is finitely generated, which is equivalent to the claim. $\square$

Proof of Theorem 4.2.1. We have to prove that if K is noetherian, then the variety $\omega \mathcal{A}$, determined by the identity (1), satisfies the descending chain condition on subvarieties. Let $F_{\mathcal{A}}$ be the free abelian group of countable rank with free generators $x_1, x_2, \ldots$. Then it is clear that the desired property is equivalent to the ascending chain condition on completely invariant ideals of the group algebra $K F_{\mathcal{A}}$.

First we note that the ring of polynomials $K[X]$ is naturally contained in $K F_{\mathcal{A}}$, and hence for each completely invariant ideal I of $K F_{\mathcal{A}}$ the intersection $I \cap K[X]$ is an ideal of $K[X]$. Evidently it is a Φ-ideal. Using the commutativity of $F_{\mathcal{A}}$, it is easy to see that the ideal of $K F_{\mathcal{A}}$ generated by $I \cap K[X]$ is just I. Consequently

$$I \mapsto I \cap K[X]$$

is a one-to-one map of the set of completely invariant ideals of $K F_{\mathcal{A}}$ into the set of Φ-ideals of $K[X]$. By Theorem 4.2.4, the latter satisfies the ascending chain condition. Since the map (7) preserves inclusions, the former satisfies this condition as well. $\square$

Cohen [10] used Theorem 4.2.5 to prove that every metabelian variety of groups is finitely based. Historically it was the first application of the method of partially well-ordered sets to the finite basis problem. Since then the method has been substantially developed and applied in a number of papers (see for instance [8', 65, 89]). Probably the most sophisticated application of this technique is due to Krasil'nikov [47] who recently announced several important results. One of these results provides a far-going generalization of Theorem 4.2.1.

4.2.6. Theorem. *Over an arbitrary noetherian ring K, the variety of group representations defined by the identity*

$$(z_1 z_2 - z_2 z_1)(z_3 z_4 - z_4 z_3) \ldots (z_{2n-1} z_{2n} - z_{2n} z_{2n-1}) \tag{7}$$

is Specht.

4.2.7. Corollary. *Every triangulable representation over a noetherian ring has a finite basis of identities.* □

When this book was written, the proof of Theorem 4.2.6 had been published only for K a field of characteristic 0 or $p > n$ [48].[1]

It is clear that a representation $\rho = (V, G)$ satisfies the identity (10) if and only if it is *stable-by-abelian*, that is, if G has a normal subgroup H acting stably on V and such that G/H is abelian. Thus Theorem 4.2.6 can be reformulated as follows: every stable-by-abelian representation is finitely based. On the other hand, by Theorem 3.4.3 every stable-by-finite representation is finitely based. This naturally suggests the following question.

4.2.8. Problem. *Let $\rho = (V, G)$ be a representation over a noetherian ring K, and let $N \triangleleft H \triangleleft G$ with (V, N) stable, H/N abelian and G/H finite (in other words, ρ is stable-by-abelian-by-finite). Will the variety var ρ be Specht, or at least finitely based? In particular, is it true if K is a field?*

A positive solution to this problem would contain both Theorems 3.4.3 and 4.2.6, and would also give a positive answer to the following important particular case of Problem 3.5.7.

---

[1] Recently we received preprints containing the proof in the general case.

4.2.9. Problem. *Does every finite-dimensional representation of a soluble group over a field have a finite basis of identities?*

Indeed, if $\rho = (V, G)$ is a finite-dimensional representation of a soluble group G, then without loss of generality G may be regarded as a soluble matrix group. By the Kolchin–Mal'cev theorem [62], G is triangulable-by-finite whence ρ is stable-by-abelian-by-finite.

We apply now the results of the last two sections to the study of multilinear identities of the form

$$z_1 z_2 \ldots z_n - z_{\sigma(1)} z_{\sigma(2)} \cdots z_{\sigma(n)} \tag{8}$$

with σ a nontrivial permutation on n letters. They are called *permutation identities*. Every set of permutation identities of algebras over any ring K is finitely based (this follows, for instance, from [81]). Together with Corollary 4.1.6, this implies

4.2.10. Corollary. *Every set of permutation identities of group representations over an arbitrary ring is finitely based.* □

One can now naturally ask: *does every permutation identity determine a Specht variety* (provided, of course, that the ground ring is noetherian)? Similar questions on permutation identities of rings, semigroups, etc. were quite popular and, under certain restrictions on the ground ring and the identity (1), were solved in a number of papers (see for instance [52, 65, 81, 89]). For group representations the answer turns out to be positive without any restrictions.

4.2.11. Theorem. *Every permutation identity of group representations over an arbitrary noetherian ring determines a Specht variety.*

This result contains, of course, Theorem 4.2.1 as a very particular case. In proving it we will need one well known fact on polynomial identities. Recall that $K\langle Z \rangle$ is the subalgebra of KF generated by the variables $z_i = x_i - 1$. It is a free associative algebra and the z_i are its free generators.

4.2.12. Lemma (Latyshev [52]). *The T-ideal of $K\langle Z\rangle$ generated by (1) contains the identity*

$$z_1 \ldots z_k [z', z''] z_{k+1} \ldots z_n$$

where $[z', z''] = z'z'' - z''z'$ and $k+1$ is the first number with $\sigma(k+1) \neq k+1$.

Proof. Let $\equiv$ denote the equality in $K\langle Z\rangle$ modulo the T-ideal generated by (8). Then (8) can be rewritten as

$$z_1 \ldots z_k z_{k+1} \ldots z_n \equiv z_1 \ldots z_k z_{\sigma(k+1)} \cdots z_{\sigma(n)}. \tag{9}$$

Applying to (9) the endomorphism $z_k \mapsto z_k z$ of $K\langle Z\rangle$ (or multiplying (9) by z on the left if $k = 0$), we have

$$z_1 \ldots z_k z z_{k+1} \ldots z_n \equiv z_1 \ldots z_k z z_{\sigma(k+1)} \cdots z_{\sigma(n)}. \tag{10}$$

On the other hand, the endomorphism $z_{\sigma(k+1)} \mapsto z z_{\sigma(k+1)}$ of $K\langle Z\rangle$, applied to (9), yields

$$z_1 \ldots z_k z_{k+1} \ldots z z_{\sigma(k+1)} \ldots z_n \equiv z_1 \ldots z_k z z_{\sigma(k+1)} \cdots z_{\sigma(n)}. \tag{11}$$

Comparing (10) and (11), we have

$$z_1 \ldots z_k z z_{k+1} \ldots z_n \equiv z_1 \ldots z_k z_{k+1} \ldots z z \sigma(k+1) \ldots z_{\sigma(n)}. \tag{12}$$

Applying the endomorphisms $\{z \mapsto z',\ z_{k+1} \mapsto z'' z_{k+1}\}$, $\{z \mapsto z'',\ z_{\sigma(k+1)} \mapsto z' z_{\sigma(k+1)}\}$ and $\{z \mapsto z'' z'\}$ to (12), we obtain respectively:

$$z_1 \ldots z_k z' z'' z_{k+1} \ldots z_n \equiv z_1 \ldots z_k z'' z_{k+1} \ldots z' z_{\sigma(k+1)} \cdots z_{\sigma(n)}, \tag{13}$$

$$z_1 \ldots z_k z'' z_{k+1} \ldots z' z_{\sigma(k+1)} \cdots z_n \equiv$$
$$\equiv z_1 \ldots z_k z_{k+1} \ldots z'' z' z_{\sigma(k+1)} \cdots z_{\sigma(n)}, \tag{14}$$

$$z_1 \ldots z_k z'' z' z_{k+1} \ldots z_n \equiv z_1 \ldots z_k z_{k+1} \ldots z'' z' z_{\sigma(k+1)} \cdots z_{\sigma(n)}. \tag{15}$$

It follows from (13)–(15) that

$$z_1 \ldots z_k z' z'' z_{k+1} \ldots z_n \equiv z_1 \ldots z_k z'' z' z_{k+1} \ldots z_n$$

which is equivalent to the claim. $\square$

Proof of Theorem 4.2.11. Let I be the completely invariant ideal of KF generated by (8). Denoting by $\equiv$ the equality in KF modulo I, and using Lemma 4.2.12 and Corollary 4.1.5, we obtain

$$z_1 z_2 \ldots z_k [z', z''] z_{k+1} \ldots z_n \equiv 0.$$

But then

$$z_2 z_1 \ldots z_k [z', z''] z_{k+1} \ldots z_n \equiv 0,$$

and hence

$$[z_1, z_2] z_3 \ldots z_k [z', z''] z_{k+1} \ldots z_n \equiv 0.$$

Repeating this argument, one can replace here $z_3 z_4$ by $[z_3, z_4]$ and so on. In the end we get

$$[z_1, z_2][z_3, z_4] \ldots [z_{2m-1}, z_{2m}] \equiv 0$$

where the exact value of m depends on the evenness or oddness of k and $n-k$, but in any case does not exceed $[\frac{n}{2}] + 2$. It remains to apply Theorem 4.2.6. $\square$

A weaker form of this result has been proved in [102].

4.3. Multiplication of pure varieties

We know from § 0.3, that the set $\mathrm{M}(K)$ of varieties of group representations over a given K is a semigroup with respect to a naturally defined multiplication. If K is a field, the abstract structure of this semigroup was described in [79]: $\mathrm{M}(K)$ *is a free semigroup with 0 and 1*. This result is analogous to the well known Neumanns–Shmel'kin Theorem [68, 23.32] on the semigroup of varieties of groups and was actually inspired by the latter. Since $\mathrm{M}(K)$ is anti-isomorphic to the semigroup of verbal ideals of KF, it can be reformulated as follows: *if K is a field then the semigroup of verbal ideals of KF is free.* A more general result was later obtained by Bergman and Lewin [6].

If the ground ring is not a field, our knowledge of the semigroup of varieties is rather poor. First, this semigroup is not free. Indeed, let R be a commutative ring with unit but not a field. Take any ideal $\mathfrak{a}$ of R and consider the ideal $\mathfrak{a}F$ of the group algebra RF. Clearly $\mathfrak{a}F$ is a verbal ideal permutable with each ideal of RF and so $\mathrm{M}(R)$ has a nontrivial center. Moreover, the map $\mathfrak{a} \mapsto \mathfrak{a}F$ is a monomorphism of the semigroup of ideals of R into the semigroup of verbal ideals of RF. Therefore one should expect to obtain a characterization of the semigroup $\mathrm{M}(R)$ only if the ideal structure of R is simple enough.

From now on R is an integral domain and K is its field of fractions. It is natural that the first step in the study of the semigroup $\mathrm{M}(R)$ consists in understanding the relationship between $\mathrm{M}(R)$ and $\mathrm{M}(K)$. The following notion proves to be useful here. A representation $\rho : G \to \mathrm{Aut}_R V$ over R is said to be *pure* if V is a torsion-free R-module; a variety of group representations over R is pure if it is generated by pure representations.

4.3.1. Lemma. *The following properties are equivalent:*
  (i) *$\mathcal{X}$ is a pure variety;*
  (ii) *all free representations of $\mathcal{X}$ are pure;*
  (iii) *$RF/\mathrm{Id}\,\mathcal{X}$ is a torsion-free R-module.*

P r o o f. If $\mathcal{X}$ is generated by pure representations, then by Lemma 0.2.6 its free representations are pure. On the other hand, $\mathcal{X}$ is generated by $\mathrm{Fr}\,\mathcal{X} = (RF/\mathrm{Id}\,\mathcal{X}, F)$ and so (iii) implies (i). $\square$

4.3.2. Lemma. *Let $\nu : \mathrm{M}(R) \to \mathrm{M}(K)$ and $\nu' : \mathrm{M}(K) \to \mathrm{M}(R)$ be the maps defined in § 0.4. Then*
  (i) *$\mathcal{Y}^{\nu'}$ is pure for every $\mathcal{Y} \in \mathrm{M}(K)$;*
  (ii) *$\mathcal{Y}^{\nu'\nu} = \mathcal{Y}$ for every $\mathcal{Y} \in \mathrm{M}(K)$;*
  (iii) *$\mathcal{X}^{\nu\nu'} = \mathcal{X}$ if and only if $\mathcal{X}$ is pure.*

P r o o f. (i) is trivial. To prove (ii), recall that $\mathrm{Id}(\mathcal{X}^\nu) = K(\mathrm{Id}\,\mathcal{X})$ and $\mathrm{Id}(\mathcal{Y}^{\nu'}) = \mathrm{Id}\,\mathcal{Y} \cap RF$. Therefore we have to show that if J is a verbal ideal in KF, then $K(J \cap RF) = J$. Take any $u \in J$ and let λ be the product of the denominators of all its coefficients. Then $\lambda u \in J \cap RF$ and

$u = \lambda^{-1} \cdot \lambda u \in K(J \cap RF)$, whence $J = K(J \cap RF)$. The proof of (iii) is also straightforward. $\square$

According to 0.4.1, ν is always a homomorphism. Hence, by Lemma 4.3.2, we have an epimorphism of semigroups $\nu : \mathrm{M}(R) \to \mathrm{M}(K)$ and an injective map $\nu' : \mathrm{M}(K) \to \mathrm{M}(R)$ giving a one-to-one correspondence between *all* varieties over K and *pure* varieties over R. The following question suggests itself: *Is the map ν' a homomorphism of semigroups? In other words, is it true that pure varieties form a subsemigroup in* $\mathrm{M}(R)$*?* This problem, raised by Plotkin in the early 1970's, can also be reformulated in terms of verbal ideals: *is it true that if R is an integral domain and K its field of fractions, then*

$$IJ \cap RF = (I \cap RF)(J \cap RF) \qquad (1)$$

for any verbal ideals I and J of KF?

If R is a Dedekind domain, a positive solution of the problem was obtained by the author [94, Corollary 9.9]. This immediately led to a natural ring-theoretic question: for which integral domains R and ideals I and J of KF (not necessarily verbal!) is the equality (1) true? Answering this question, Bergman proved the following fact.

4.3.3. Theorem (Bergman [5']). *If R is a Dedekind domain then (1) is true for every right ideal I and every left ideal J of KF.*

The aim of this section is two-fold. First, we will present a complete proof of Theorem 4.3.3 (with the kind permission of its author). Second, we will sketch a counterexample showing that over an *arbitrary* domain the product of pure varieties need not be pure. This result is quite recent and its detailed exposition will appear elsewhere.

Recall first a few standard notions of commutative algebra (see for example [104, Chapter 5]). For an arbitrary domain R, an R-submodule M of its field of fractions K is called a *fractional ideal of R* if $Md \subseteq R$ for some $0 \neq d \in R$. Clearly M is a fractional ideal if and only if $M = \mathfrak{a}d^{-1}$ for some ideal $\mathfrak{a}$ of R and $0 \neq d \in R$. In particular, the "ordinary" ideals of R are fractional ideals; they will be always denoted by lowercase Gothic letters.

If M and N are fractional ideals, then their product $MN = \{\sum a_i b_i \mid a_i \in M, b_i \in N\}$ is also a fractional ideal. Therefore the set of all fractional ideals is a monoid with the identity element R. A fractional ideal M is *invertible* if there exists a fractional ideal $N = M^{-1}$ such that $MN = R$. If M is invertible, its inverse M^{-1} coincides with the set $(R : M) = \{x \in K \mid Mx \subseteq R\}$. We emphasize that if an ideal $\mathfrak{a}$ of R is invertible, then its inverse $\mathfrak{a}^{-1}$ is a fractional ideal but not an ideal, unless $\mathfrak{a} = R$.

4.3.4. Lemma. *If $\mathfrak{a}$ and $\mathfrak{b}$ are ideals of R such that $\mathfrak{a}, \mathfrak{b}, \mathfrak{a} + \mathfrak{b}$ are invertible, then $\mathfrak{a} \cap \mathfrak{b}$ is invertible and $(\mathfrak{a} \cap \mathfrak{b})^{-1} = \mathfrak{a}^{-1} + \mathfrak{b}^{-1}$.*

P r o o f. Clearly $\mathfrak{a}^{-1} + \mathfrak{b}^{-1} = \mathfrak{a}^{-1}\mathfrak{b}^{-1}(\mathfrak{a} + \mathfrak{b})$. Being a product of three invertible ideals, $\mathfrak{a}^{-1} + \mathfrak{b}^{-1}$ is invertible. Its inverse must be $\{x \mid x(\mathfrak{a}^{-1} + \mathfrak{b}^{-1}) \subseteq R\} = \mathfrak{a} \cap \mathfrak{b}$. $\square$

4.3.5. Lemma. *Let $V \subseteq W$ be vector spaces over K and M an R-submodule of W. Then:*

(i) *if $\mathfrak{a}$ is an invertible ideal of R, then $V \cap \mathfrak{a}M = \mathfrak{a}(V \cap M)$;*

(ii) *if $\mathfrak{a}, \mathfrak{b}, \mathfrak{a} + \mathfrak{b}$ are invertible ideals of R, then $\mathfrak{a}M \cap \mathfrak{b}M = (\mathfrak{a} \cap \mathfrak{b})M$.*

P r o o f. (i) Clearly $V \cap \mathfrak{a}M \supseteq \mathfrak{a}(V \cap M)$. On the other hand, $\mathfrak{a}^{-1}(V \cap \mathfrak{a}M) \subseteq V \cap \mathfrak{a}^{-1}\mathfrak{a}M = V \cap M$. Multiplying the second inclusion by $\mathfrak{a}$, we obtain $V \cap \mathfrak{a}M \subseteq \mathfrak{a}(V \cap M)$.

(ii) Note that $(\mathfrak{a}^{-1} + \mathfrak{b}^{-1})(\mathfrak{a}M \cap \mathfrak{b}M) \subseteq M$. Multiplying by $\mathfrak{a} \cap \mathfrak{b}$ and applying the previous lemma, we get $\mathfrak{a}M \cap \mathfrak{b}M \subseteq (\mathfrak{a} \cap \mathfrak{b})M$. The reverse inclusion is obvious. $\square$

4.3.6. Lemma. *Let $\mathfrak{p}$ be a maximal ideal of R, $R_{\mathfrak{p}}$ the localization of R at $\mathfrak{p}$, and i a positive integer. Then the natural homomorphism $\alpha : R/\mathfrak{p}^i \to R_{\mathfrak{p}}/\mathfrak{p}^i R_{\mathfrak{p}}$ is an isomorphism.*

P r o o f. We first show that every element a of R not in $\mathfrak{p}$ has an invertible image in $R/\mathfrak{p}^i$. Since $R/\mathfrak{p}$ is a field, there exists $b \in R$ with $ab - 1 \in \mathfrak{p}$.

Then $(ab-1)^i(-1)^{i+1} \in \mathfrak{p}^i$, but it is evident that $(ab-1)^i(-1)^{i+1} = ac-1$ for some $c \in R$. Thus $ac-1 \in \mathfrak{p}^i$ whence the claim follows.

To prove that the homomorphism α is surjective, take any $a/b + \mathfrak{p}^i R_{\mathfrak{p}} \in R_{\mathfrak{p}}/\mathfrak{p}^i R_{\mathfrak{p}}$. Here $b \notin \mathfrak{p}$, and by the previous remark there exists c such that $bc - 1 \in \mathfrak{p}^i$. But then for $ac + \mathfrak{p}^i \in R/\mathfrak{p}^i$ we have

$$(ac + \mathfrak{p}^i)^\alpha = ac + \mathfrak{p}^i R_{\mathfrak{p}} = \frac{a}{b} + \mathfrak{p}^i R_{\mathfrak{p}}.$$

Finally, let $a - b \in \mathfrak{p}^i R_{\mathfrak{p}}$, then $a - b = c/d$ where $c \in \mathfrak{p}^i, d \notin \mathfrak{p}$. Therefore $(a-b)d \in \mathfrak{p}^i$ and since d is invertible modulo $\mathfrak{p}^i$, we have $a - b \in \mathfrak{p}^i$. This shows that α is injective. $\square$

4.3.7. Lemma. *Let W, M, $\mathfrak{p}$, i be as in the previous lemmas. Then*

$$\mathfrak{p}^i R_{\mathfrak{p}} M \cap M = \mathfrak{p}^i M \quad and \quad M/\mathfrak{p}^i M \simeq R_{\mathfrak{p}} M/\mathfrak{p}^i R_{\mathfrak{p}} M.$$

P r o o f. Let $x \in \mathfrak{p}^i R_{\mathfrak{p}} M \cap M$. Since $x \in \mathfrak{p}^i R_{\mathfrak{p}} M$, we have $x \in \mathfrak{p}^i a^{-1} M$ for some $a \notin \mathfrak{p}$, whence $ax \in \mathfrak{p}^i M$. As in the proof of the previous lemma, one can find $c \in R$ such that $ac - 1 \in \mathfrak{p}^i$. Together with the inclusion $x \in M$, this implies that $(ac - 1)x \in \mathfrak{p}^i M$. On the other hand, $ax \in \mathfrak{p}^i M$ and so $axc \in \mathfrak{p}^i M$. Subtracting, we get $x \in \mathfrak{p}^i M$. Thus $\mathfrak{p}^i R_{\mathfrak{p}} M \cap M \subseteq \mathfrak{p}^i M$ while the reverse inclusion is obvious.

The proof of the second assertion is similar. $\square$

Recall that a *Dedekind domain* is an integral domain R all of whose nonzero fractional ideals are invertible. Since every fractional ideal can be presented as $\mathfrak{a}d^{-1}$, where $\mathfrak{a}$ is an ideal and $0 \neq d \in R$, this is equivalent to saying that all nonzero ideals of R are invertible. Every nonzero ideal of a Dedekind domain R has a unique factorization as a product of prime ideals, every nonzero prime ideal $\mathfrak{p}$ is maximal, and the localization $R_{\mathfrak{p}}$ is a discrete valuation ring. For the details we refer to [104] or to any other textbook on commutative algebra.

Let now A be an arbitrary ring with 1, not necessary commutative. It is called a *semifir* if every finitely generated right ideal of A is a free A-module of unique rank (this condition is left-right symmetric). Let $\phi : A \to \bar{A}$ be a surjective homomorphism of rings. It induces the homomorphism of general linear groups $\phi^* : GL_m(A) \to GL_m(\bar{A})$ which, in general, is not surjective. We will say that ϕ^* is *almost surjective* if for each matrix $\bar{U} \in GL_m(\bar{A})$ one can find a diagonal matrix $\bar{D} \in GL_m(\bar{A})$ such that there exists $U \in GL_m(A)$ with $U\phi^* = \bar{U}\bar{D}$.

4.3.8. Lemma. *Let R be a Dedekind domain, K its field of fractions, $\mathfrak{p}$ a nonzero prime ideal of R. Let B be a K-algebra and A an R-subalgebra of B such that*

(i) *$A/\mathfrak{p}A$ is a semifir;*

(ii) *for every $m > 0$ the homomorphism $GL_m(A) \to GL_m(A/\mathfrak{p}A)$ is almost surjective.*

Then for every right ideal I of B, left ideal J of B and ideal $\mathfrak{a}$ of R

$$\mathfrak{a}(I \cap A)(J \cap A) \cap \mathfrak{a}\mathfrak{p}A = \mathfrak{a}\mathfrak{p}(I \cap A)(J \cap A).$$

Proof. "$\supseteq$" is evident. To prove the reverse inclusion, note that $R_\mathfrak{p}$ is a discrete valuation ring. Its maximal ideal $\mathfrak{p}R_\mathfrak{p}$ is principal, so $\mathfrak{p}R_\mathfrak{p} = tR_\mathfrak{p}$ for some $t \in R_\mathfrak{p}$. The ideal $\mathfrak{a}R_\mathfrak{p}$ of $R_\mathfrak{p}$ must be a power of $\mathfrak{p}R_\mathfrak{p}$, hence $\mathfrak{a}R_\mathfrak{p} = \mathfrak{p}^n R_\mathfrak{p} = t^n R_\mathfrak{p}$ for some $n \geq 0$.

Take an arbitrary $z \in \mathfrak{a}(I \cap A)(J \cap A) \cap \mathfrak{a}\mathfrak{p}A$. Since $z \in \mathfrak{a}(I \cap A)(J \cap A)$, we can write

$$z = \sum_{i=1}^{m} x_i y_i \qquad (x_i \in \mathfrak{a}(I \cap A),\ y_i \in J \cap A). \tag{2}$$

For each i we have $x_i \in \mathfrak{a}A \subseteq \mathfrak{a}R_\mathfrak{p}A = t^n R_\mathfrak{p}A$, and so $x_i = t^n x_i'$ for some $x_i' \in R_\mathfrak{p}A$. On the other hand, since $z \in \mathfrak{a}\mathfrak{p}A \subseteq \mathfrak{a}\mathfrak{p}R_\mathfrak{p}A = t^{n+1}R_\mathfrak{p}A$, it follows that $z = t^{n+1}z'$ for some $z' \in R_\mathfrak{p}A$. If we now divide (2) by t^n and reduce modulo $\mathfrak{p}R_\mathfrak{p}A$, we get

$$0 = \sum_{i=1}^{m} \bar{x}_i'\bar{y} \quad \text{in} \quad R_\mathfrak{p}A/\mathfrak{p}R_\mathfrak{p}A = \bar{A}. \tag{3}$$

In view of Lemma 4.3.7, $\bar{A} \simeq A/\mathfrak{p}A$ and therefore $\bar{A}$ is a semifir. By [13, Theorem 1.1.1] the equation (3) can be trivialized by some matrix $\bar{U} \in GL_m(\bar{A})$. This means that in the relation

$$(\bar{x}'\bar{U}) \cdot (\bar{U}^{-1}\bar{y}) = 0, \tag{4}$$

where $\bar{x}' = (\bar{x}'_1, \ldots, \bar{x}'_m)$ and $\bar{y} = (\bar{y}_1, \ldots, \bar{y}_m)^T$, for each i either the i-th component of $\bar{x}'\bar{U}$ is zero or the i-th component of $\bar{U}^{-1}\bar{y}$ is zero. By (ii), one can find a diagonal matrix $\bar{D} \in GL_m(\bar{A})$ such that there exists $U \in GL_m(A)$ whose image in $GL_m(\bar{A})$ equals $\bar{U}\bar{D}$. Then we first have

$$(\bar{x}'\bar{U}\bar{D}) \cdot (\bar{D}^{-1}\bar{U}\bar{y}) = 0$$

where again for each i either the i-th component of $\bar{x}'\bar{U}\bar{D}$ is zero or the i-th component of $\bar{D}^{-1}\bar{U}^{-1}\bar{y}$ is zero. Second, applying U to (2), we obtain

$$z = \sum x_i y_i = x \cdot y = (xU) \cdot (U^{-1}y) = \hat{x} \cdot \hat{y} = \sum \hat{x}_i \hat{y}_i \qquad (5)$$

where we still have $\hat{x}_i \in \mathfrak{a}(I \cap A)$, $\hat{y}_i \in J \cap A$, $\hat{x}_i = t^n \hat{x}'_i$, $\hat{x}'_i \in R_\mathfrak{p} A$. In addition, now for each i

$$\hat{x}'_i \in \mathfrak{p} R_\mathfrak{p} A \quad \text{or} \quad \hat{y}_i \in \mathfrak{p} R_\mathfrak{p} A, \qquad (6)$$

because our construction guarantees that the homomorphism $R_\mathfrak{p} A \to \bar{A}$ maps $\hat{x}'_i$ to the i-th component of $\bar{x}'\bar{U}\bar{D}$ and $\hat{y}_i$ to the i-th component of $\bar{D}^{-1}\bar{U}^{-1}\bar{y}$. To complete the proof, it suffices to show that each summand $\hat{x}_i \hat{y}_i$ of (5) belongs to $\mathfrak{a}\mathfrak{p}(I \cap A)(J \cap A)$.

Choose any value of i for which the second alternative in (6) holds. Keeping in mind that A is flat as an R-module and using successively Lemmas 4.3.7 and 4.3.5(ii), we have

$$\hat{y}_i \in \mathfrak{p} R_\mathfrak{p} A \cap (J \cap A) = (\mathfrak{p} R_\mathfrak{p} A \cap A) \cap J = \mathfrak{p} A \cap J = \mathfrak{p}(J \cap A)$$

whence $\hat{x}_i \hat{y}_i \in \mathfrak{a}(I \cap A)\mathfrak{p}(J \cap A) = \mathfrak{a}\mathfrak{p}(I \cap A)(J \cap A)$, as desired.

Now suppose the first alternative in (6) holds. Then $\hat{x}_i = t^n \hat{x}'_i \in \mathfrak{p}^{n+1} R_\mathfrak{p} A$, so this element lies in $\mathfrak{p}^{n+1} R_\mathfrak{p} A \cap A = \mathfrak{p}^{n+1}A$ (Lemma 4.3.7). Since it also lies in $\mathfrak{a}A$, by Lemma 4.3.5(ii) it belongs to $(\mathfrak{a} \cap \mathfrak{p}^{n+1})A$. Furthermore, $\hat{x}_i \in I$ and by Lemma 4.3.5(i) we have

$$\hat{x}_i \in I \cap (\mathfrak{a} \cap \mathfrak{p}^{n+1})A = (\mathfrak{a} \cap \mathfrak{p}^{n+1})(I \cap A). \qquad (7)$$

But it is easy to show that $\mathfrak{a} \cap \mathfrak{p}^{n+1} = \mathfrak{a}\mathfrak{p}$. Indeed, from $\mathfrak{a}R_\mathfrak{p} = \mathfrak{p}^n R_\mathfrak{p}$ we deduce that $\mathfrak{a}\mathfrak{p} \subseteq \mathfrak{p}^{n+1}$, and so

$$\mathfrak{a}\mathfrak{p} \subseteq \mathfrak{a} \cap \mathfrak{p}^{n+1} \subset \mathfrak{a}.$$

The second inclusion here is proper, for otherwise $\mathfrak{p}^{n+1} \supseteq \mathfrak{a}$ which is in contradiction with $\mathfrak{a}R_\mathfrak{p} = \mathfrak{p}^n R_\mathfrak{p}$. But by the unique factorization of ideals in a Dedekind domain, there can be no ideals strictly between $\mathfrak{a}$ and $\mathfrak{a}\mathfrak{p}$. Thus $\mathfrak{a} \cap \mathfrak{p}^{n+1} = \mathfrak{a}\mathfrak{p}$. By (7), $\hat{x}_i \in \mathfrak{a}\mathfrak{p}(I \cap A)$ and so $\hat{x}_i \hat{y}_i \in \mathfrak{a}\mathfrak{p}(I \cap A)(J \cap A)$. $\square$

4.3.9. Lemma. *Under conditions and notation of the previous lemma, let (i) and (ii) be satisfied for every prime ideal of R. Then*

$$(I \cap A)(J \cap A) \cap \mathfrak{a}A = \mathfrak{a}(I \cap A)(J \cap A).$$

Proof. Let $\mathfrak{a} = \mathfrak{p}_1 \ldots \mathfrak{p}_n$ be the prime factorization of $\mathfrak{a}$. We proceed by induction on n. If $n = 1$, then the previous lemma (with $\mathfrak{a} = R$) gives the result. Assume that for $n - 1$ everything has been proved. By Lemma 4.3.8

$$\mathfrak{p}_1 \ldots \mathfrak{p}_{n-1}(I \cap A)(J \cap A) \cap \mathfrak{p}_1 \ldots \mathfrak{p}_n A = \mathfrak{p}_1 \ldots \mathfrak{p}_n(I \cap A)(J \cap A).$$

On the other hand, by induction hypothesis

$$(I \cap A)(J \cap A) \cap \mathfrak{p}_1 \ldots \mathfrak{p}_{n-1} A = \mathfrak{p}_1 \ldots \mathfrak{p}_{n-1}(I \cap A)(J \cap A).$$

Therefore

$$\begin{aligned}
(I \cap A)(J \cap A) \cap \mathfrak{a}A &= (I \cap A)(J \cap A) \cap \mathfrak{p}_1 \ldots \mathfrak{p}_n A \\
&= (I \cap A)(J \cap A) \cap \mathfrak{p}_1 \ldots \mathfrak{p}_{n-1} A \cap \mathfrak{p}_1 \ldots \mathfrak{p}_n A \\
&= \mathfrak{p}_1 \ldots \mathfrak{p}_{n-1}(I \cap A)(J \cap A) \cap \mathfrak{p}_1 \ldots \mathfrak{p}_n A \\
&= \mathfrak{p}_1 \ldots \mathfrak{p}_n(I \cap A)(J \cap A) = \mathfrak{a}A. \quad \square
\end{aligned}$$

Proof of Theorem 4.3.3. Denote $RF = A$ and $KF = B$. We have to prove that
$$IJ \cap A = (I \cap A)(J \cap A) \tag{8}$$
for a right ideal I of B and a left ideal J of B. Note first that A satisfies the conditions (i) and (ii) of Lemma 4.3.8 for every prime ideal $\mathfrak{p}$ of R. Indeed, $A/\mathfrak{p}A$ is isomorphic to $(R/\mathfrak{p}R)F$, the group algebra of the free group F over the field $R/\mathfrak{p}R$. By [4', Corollary IV.5.17], $A/\mathfrak{p}A = \bar{A}$ is a *generalized*

euclidean ring. This first means that $\bar{A}$ is a fir, whence (i) follows. Second, this means that every matrix $\bar{U} \in GL_m(\bar{A})$ can be written as a product $\bar{U} = \bar{E}_1 \dots \bar{E}_s \bar{D}$, where each $\bar{E}_k$ is an elementary matrix (i.e. a matrix of the form $I + \bar{a} e_{ij}$ with $\bar{a} \in \bar{A}$ and $i \neq j$) and $\bar{D}$ is an invertible diagonal matrix. Then $\bar{U} \bar{D}^{-1} = \bar{E}_1 \dots \bar{E}_s$. Since every elementary matrix can be lifted from $\bar{A}$ to A, the matrix $\bar{U} \bar{D}^{-1}$ can be lifted as well, and we obtain (ii).

To prove the nontrivial part of (8), note that every $a \in IJ$ can be written in the form $a = \alpha^{-1} \sum u_i v_i$, where $0 \neq \alpha \in R$, $u_i \in I \cap A$, $v_i \in J \cap A$. If now $a \in IJ \cap A$, then

$$\alpha a \in (I \cap A)(J \cap A) \cap \alpha A.$$

On the other hand, by Lemma 4.3.9

$$(I \cap A)(J \cap A) \cap (\alpha)A = (\alpha)(I \cap A)(J \cap A).$$

Therefore $\alpha a \in \alpha(I \cap A)(J \cap A)$ and so $a \in (I \cap A)(J \cap A)$, as required. $\square$

4.3.10. Corollary (Vovsi [94, Corollary 9.9]). *Let R be a Dedekind domain and K its field of fractions. Then the pure varieties of group representations over R form a subsemigroup in $\mathbb{M}(R)$ which is isomorphic to $\mathbb{M}(K)$ (and therefore is free).* $\square$

The question whether this fact is valid over an arbitrary integral domain has remained open for more than fifteen years. Quite recently the author has found a counterexample showing that it is not the case.

4.3.11. Theorem. *There exists an integral domain R and a verbal ideal I of KF, where K is the field of fractions of R, such that*

$$I^2 \cap RF \neq (I \cap RF)^2. \tag{9}$$

S k e t c h   o f   P r o o f. Let k be a field of characteristic $\neq 2, 3$, $K = k(s)$ be the field of rational functions over k in a variable s, and $R = k[s^2, s^3]$. Clearly K is the field of fractions of R. Let I be the verbal ideal of KF generated by two elements

$$u = z_1^2 z_2 - s z_2 z_1^2, \quad v = z_1 z_2 z_3 z_4$$

where, as usual, $z_i = x_i - 1$. To prove (9), consider the element

$$w = (su)^2 = s^2 z_1^2 z_2 z_1^2 z_2 - s^3 z_1^2 z_2^2 z_1^2 - s^3 z_2 z_1^4 z_2 + s^4 z_2 z_1^2 z_2 z_1^2$$

in KF. It belongs to I^2, and since its coefficients lie in R, we have $w \in I^2 \cap RF$. It remains to show that $w \notin (I \cap RF)^2$.

Suppose the contrary. Then

$$w = u_1 v_1 + \cdots + u_n v_n \qquad (10)$$

where $u_i, v_i \in I \cap RF$. Since $\Delta^3 \supset I \supset \Delta^4$, the ideal I is homogeneous. Using this observation and some arguments from § 1.2, one can reduce (10) to the case when all the u_i, v_i involve only the variables z_1, z_2 and are homogeneous of degree 2 in z_1 and homogeneous of degree 1 in z_2. Thus we can assume that F is the free group of rank 2. Then a careful analysis of the module KF/I shows that $\dim_K(KF/I) = 9$. Since $\dim_K(KF/\Delta^4) = 15$, it follows that $\dim_K(I/\Delta^4) = 6$. Furthermore, one can show that the elements

$$z_1^3, \quad z_2^3, \quad z_1^2 z_2 - s z_2 z_1^2, \quad z_2^2 z_1 - s z_1 z_2^2,$$

$$z_1 z_2 z_1 + (1+s) z_2 z_1^2, \quad z_2 z_1 z_2 + (1+s) z_1 z_2^2$$

form a basis of I over Δ^4. This implies that each u_i and v_i is a K-linear combination of *two* elements $z_1^2 z_2 - s z_2 z_1^2$, $z_1 z_2 z_1 + (1+s) z_2 z_1^2$. However, substituting these combinations in (10) and combining similar terms, we will eventually see that the coefficients of the resulting expression can not belong to R. This yields the desired contradiction.

In conclusion we note that the results of the present section are also valid for free associative algebras and T-ideals. Namely, let R be an integral domain, K its field of fractions, $R\langle X\rangle$ and $K\langle X\rangle$ the free associative algebras over R and K respectively. Then:

(i) If R is a Dedekind domain, then

$$IJ \cap R\langle X\rangle = (I \cap R\langle X\rangle)(J \cap R\langle X\rangle)$$

for every right ideal I and every left ideal J of $K\langle X\rangle$.

(ii) Let k be a field of characteristic $\neq 2, 3$, $K = k(s)$ and $R = k[s^2, s^3]$. If I is the T-ideal of $K\langle X\rangle$ generated by $x_1^2 x_2 - sx_2 x_1^2$ and $x_1 x_2 x_3 x_4$, then

$$I^2 \cap R\langle X\rangle \neq (I \cap R\langle X\rangle)^2.$$

The proofs of these statements are virtually the same as those of Theorem 4.3.3 and 4.3.11. Moreover, (i) *formally* follows from the proof of Theorem 4.3.3: one should only set in the latter $A = R\langle X\rangle$ and $B = K\langle X\rangle$.

4.4. Intersection of pure varieties

In the present short section we follow [96].

Let $\mathcal{X}$ and $\mathcal{Y}$ be two pure varieties over an arbitrary integral domain. We already know that if this domain is good enough, then the product $\mathcal{X}\mathcal{Y}$ is also pure. Now one can naturally ask whether the property of pureness is compatible with the lattice operations on varieties. Clearly the join $\mathcal{X} \vee \mathcal{Y}$ of pure varieties is always pure, so that it remains to understand whether *the intersection of pure varieties is pure*. At the end of the section we will show that it is not the case even over the ring of integers. To obtain this result, we will define and investigate a natural series of varieties which are of some interest in their own right. An important property of these varieties is that their identities can be completely described in terms of the Magnus presentation of the free group algebra by formal power series.

Let K be an arbitrary commutative ring with 1. For every $n = 2, 3, \ldots$ denote by $\mathcal{V}_n$ the variety of group representations over K defined by the identities

$$(x-1)^n \quad \text{and} \quad (x_1-1)(x_2-1)\ldots(x_{n+1}-1).$$

In other words, $\mathcal{V}_n = \mathcal{U}_n \cap \mathcal{S}^{n+1}$, where $\mathcal{U}_n$ is the variety of n-unipotent representations. Therefore, denoting $V_n = \operatorname{Id}\mathcal{V}_n$ and $U_n = \operatorname{Id}\mathcal{U}_n$, we have $V_n = U_n + \Delta^{n+1}$.

Our first aim is to find conditions on an element $u \in KF$ which would be necessary and sufficient for u to lie in V_n. Recall that u can be uniquely presented as formal power series

$$u = u_{(0)} + u_{(1)} + \cdots + u_{(n)} + \cdots, \quad \text{where} \quad u_{(n)} = \sum \delta_{i_1 \ldots i_n} z_{i_1} \cdots z_{i_n},$$

and that the coefficients of this decomposition are just the corresponding Fox derivatives at 1:

$$\delta_{i_1 \ldots i_n} = \partial_{i_1 \ldots i_n} u(1) \tag{1}$$

(see § 0.5). Further, if $\{i_1, \ldots, i_n\}$ is an arbitrary *multiset* (i.e. a collection of elements where some elements may occur more than once), then the set of all its permutations is denoted by $S(i_1, \ldots, i_n)$. For example, $S(1, 1, 2)$ consists of three permutations $(1, 1, 2), (1, 2, 1), (2, 1, 1)$; $S(1, 2, 3)$ consists of six permutations; etc.

The following result gives a desired criterion under a slight restriction on the ground ring K. In its formulation the indices $i, i_1, i_2, \ldots$ independently run over the set of positive integers.

4.4.1. Proposition. *The following conditions are necessary for $u \in KF$ to belong to V_n:*

(a) $u_{(0)} = u_{(1)} = \cdots = u_{(n-1)} = 0$;

(b) $\delta_{i_n \ldots i_1} = \delta_{\sigma(i_n) \ldots \sigma(i_1)}$ *for all* $\sigma \in S(i_1, \ldots, i_n)$.

If $(n-1)!$ is a unit in K, these conditions are also sufficient.

Proof. *Necessity.* Take an arbitrary $u \in V_n = U_n + \Delta^{n+1}$. Then $u \in \Delta^n$ and (a) is trivial. To prove (b), we will use (1). Since elements from Δ^{n+1} certainly satisfy (b) and since all Fox derivatives are linear maps, we may assume $u \in U_n$. It is easy to see that U_n is generated by elements $(f - 1)^n$, where $f \in F$, as a left ideal. Hence $u = \sum_k v_k (f_k - 1)^n$ for some $v_k \in KF$, $f_k \in F$. Again by the linearity of the Fox derivatives, we may assume that $u = v(f - 1)^n$ where $u \in KF$, $f \in F$.

Let $w = w(x_1, \ldots, x_n)$ be an arbitrary element from KF. For brevity, the scalar $w(1, \ldots, 1) = w(1)$ will be denoted by $\overline{w}$. Applying (4) from § 0.5, for the given $u = v(f - 1)^n$ we successively obtain:

$$\partial_{i_1} u = \partial_{i_1}(v(f-1)^n) = v(f-1)^{n-1} \partial_{i_1} f;$$

$$\partial_{i_2 i_1} u = \partial_{i_2}(v(f-1)^{n-1} \partial_{i_1} f)$$
$$= v(f-1)^{n-2} \partial_{i_2} f \cdot \overline{\partial_{i_1} f} + v(f-1)^{n-1} \partial_{i_2 i_1} f$$
$$= v[(f-1)^{n-1} \partial_{i_2 i_1} f + (f-1)^{n-2} \partial_{i_2} f \cdot \overline{\partial_{i_1} f}].$$

Assume now that for $k < n$ we have already proved that

$$\partial_{i_k \ldots i_1} u = v[(f-1)^{n-1} w_1 + (f-1)^{n-2} w_2 + \cdots + (f-1)^{n-k} w_k]$$

where $w_1, \ldots, w_k$ are some elements from KF and $w_k = \partial_{i_k} f \, \overline{\partial_{i_{k-1}} f} \ldots \overline{\partial_{i_1} f}$. Then

$$\begin{aligned}
\partial_{i_{k+1} i_k \ldots i_1} u &= \partial_{i_{k+1}} (v[(f-1)^{n-1} w_1 + \cdots + (f-1)^{n-k} w_k]) \\
&= v \partial_{i_{k+1}} [(f-1)^{n-1} w_1 + \cdots + (f-1)^{n-k} w_k] \\
&= v[(f-1)^{n-2} \partial_{i_{k+1}} f \cdot \overline{w_1} + (f-1)^{n-1} \partial_{i_{k+1}} w_1 + \\
&\quad + (f-1)^{n-3} \partial_{i_{k+1}} f \cdot \overline{w_2} + (f-1)^{n-2} \partial_{i_{k+1}} w_2 + \cdots \\
&\quad + (f-1)^{n-(k-1)} \partial_{i_{k+1}} f \cdot \overline{w_k} + (f-1)^{n-k} \partial_{i_{k+1}} w_k] \\
&= v[(f-1)^{n-1} w_1' + (f-1)^{n-2} w_2' + \cdots + (f-1)^{n-(k+1)} w_{k+1}']
\end{aligned}$$

where

$$w_{k+1}' = \partial_{i_{k+1}} f \cdot \overline{w_k} = \partial_{i_{k+1}} f \, \overline{\partial_{i_k} f} \ldots \overline{\partial_{i_1} f}.$$

By induction we conclude that

$$\partial_{i_n \ldots i_1} u = v[(f-1)^{n-1} w_1 + \cdots + (f-1) w_{n-1} + w_n]$$

where $w_n = \partial_{i_n} f \, \overline{\partial_{i_{n-1}} f} \ldots \overline{\partial_{i_1} f}$. Therefore

$$\overline{\partial_{i_n \ldots i_1} u} = \overline{v} \, \overline{\partial_{i_n} f} \, \overline{\partial_{i_{n-1}} f} \, \ldots \, \overline{\partial_{i_1} f}$$

and so $\delta_{i_n \ldots i_1} = \delta_{\sigma(i_n) \ldots \sigma(i_1)}$ for any $\sigma \in S(i_1, \ldots, i_n)$.

Sufficiency. Let $u \in KF$ satisfy (a) and (b). Then u can be written as

$$u = \sum \delta_{i_1 \ldots i_n} \left(\sum_{\sigma \in S(i_1, \ldots, i_n)} z_{\sigma(i_1)} \cdots z_{\sigma(i_n)} \right) + u'$$

where $u' \in \Delta^{n+1}$ and the first sum is taken over all multisets $\{i_1, \ldots, i_n\}$ of length n. Therefore it remains to show that

$$s(i_1, \ldots, i_n) = \sum_{\sigma \in S(i_1, \ldots, i_n)} z_{\sigma(i_1)} \cdots z_{\sigma(i_n)} \in V_n$$

for any $i_1, \ldots, i_n$.

Note first that $s(1, 2, \ldots, n)$ is just the full linearization $\lin(z^n)$ of z^n. Since $\Delta^n \supset V_n \supset \Delta^{n+1}$, the ideal V_n is homogeneous. By Lemma 1.2.2 and the subsequent note, we have

$$s(1, 2, \ldots, n) \in V_n. \tag{2}$$

Now let $\{i_1, \ldots, i_n\}$ be an arbitrary multiset. Without loss of generality we may assume that

$$i_1 = \cdots = i_{m_1} < i_{m_1+1} = \cdots = i_{m_1+m_2} < \cdots < i_{m_1+\cdots+m_{r-1}+1} = \cdots$$
$$= i_{m_1+\cdots+m_r}$$

where $m_1 + \cdots + m_r = n$. If $r = 1$, that is $i_1 = \cdots = i_n$, then $s(i_1, \ldots, i_n) = z_{i_1}^n \in V_n$ simply by the definition of V_n. Otherwise $1 \leq m_i \leq n - 1$ for each $i = 1, \ldots, r$. Let φ be an endomorphism of F such that $x_1^\varphi = x_{i_1}$, $x_2^\varphi = x_{i_2}, \ldots, x_n^\varphi = x_{i_n}$. From (2) and straightforward combinatorics we obtain

$$s(1, \ldots, n)^\varphi = m_1! \ldots m_r! s(i_1, \ldots, i_n) \in V_n.$$

Since $1 \leq m_i \leq n - 1$ and $1/(n-1)! \in K$, it follows that $s(i_1, \ldots, i_n) \in V_n$. $\square$

The established fact has several corollaries. Let A be a K-module and B its submodule. If A/B is a free K-module with a basis $\{a_i + B\}$, then we say that $\{a_i\}$ is a basis of A over B.

4.4.2. Corollary. *If $(n-1)!$ is a unit in K, then KF/V_n is a free K-module. A basis of KF over V_n consists of the elements*

$$1, z_i, z_i z_j, \ldots, z_{i_1} \ldots z_{i_n} \tag{3}$$

and also of all elements of the form $z_{\sigma(i_1)} \ldots z_{\sigma(i_n)}$, where $\{i_1, \ldots, i_n\}$ is an n-multiset and σ is its nontrivial permutation.

P r o o f. Every $u \in KF$ is uniquely presented in the form

$$u = u_{(0)} + u_{(1)} + \cdots + u_{(n)} + u'$$

where $u_{(i)}$ is its homogeneous component of degree i and $u' \in \Delta^{n+1}$. This determines a natural graded algebra structure on KF:

$$KF = A_0 \oplus A_1 \oplus \cdots \oplus A_n \oplus \Delta^{n+1}.$$

Note that $A_i \oplus \cdots \oplus A_n \oplus \Delta^{n+1} = \Delta^i$ for each $i = 1, \ldots, n$. Since V_n lies between Δ^n and Δ^{n+1}, we have $V_n = B_n \oplus \Delta^{n+1}$ for some submodule B_n of A_n. Thus

$$KF_n/V_n \simeq A_0 \oplus A_1 \oplus \cdots \oplus A_{n-1} \oplus A_n/B_n.$$

Elements (3) form a basis of $A_0 \oplus A_1 \oplus \cdots \oplus A_{n-1}$, so it suffices to show that the remaining elements from the formulation of the lemma form a basis of A_n over B_n.

Note that the module A_n is multigraded: $A_n = \oplus A_{\{i_1, \ldots, i_n\}}$ where the sum is taken over all n-multisets $\{i_1, \ldots, i_n\}$ and $A_{\{i_1, \ldots, i_n\}}$ is a free module with the basis $\{z_{\sigma(i_1)} \cdots z_{\sigma(i_n)} \mid \sigma \in S(i_1, \ldots, i_n)\}$. By Theorem 4.4.1, B_n is a multigraded submodule of A_n. More exactly, $B_n = \oplus B_{\{i_1, \ldots, i_n\}}$ where each $B_{\{i_1, \ldots, i_n\}}$ is the cyclic K-submodule of $A_{\{i_1, \ldots, i_n\}}$ generated by the element $s(i_1, \ldots, i_n)$. We have

$$A_n/B_n = \bigoplus_{\{i_1, \ldots, i_n\}} \left(A_{\{i_1, \ldots, i_n\}} / B_{\{i_1, \ldots, i_n\}} \right)$$

and it is clear that $\{z_{\sigma(i_1)} \cdots z_{\sigma(i_n)} \mid 1 \neq \sigma \in S(i_1, \ldots, i_n)\}$ is a basis of $A_{\{i_1, \ldots, i_n\}}$ over $B_{\{i_1, \ldots, i_n\}}$. $\square$

Now we can answer a question raised at the beginning of this section.

4.4.3. Corollary. *There exist pure varieties of group representations over the ring of integers whose intersection is not pure.*

Proof. Consider two varieties over $\mathbb{Z}$: $\mathcal{V}_2$ and $\omega\mathfrak{B}_2$, where $\mathfrak{B}_2$ is the variety of groups of exponent 2. By the preceding lemma, $\mathcal{V}_2$ is pure over any K. The variety $\omega\mathfrak{B}_2$ is pure by trivial reasons. We show that their intersection is not pure. If $\mathrm{Id}(\omega\mathfrak{B}_2) = I$, then $\mathrm{Id}(\mathcal{V}_2 \cap \omega\mathfrak{B}_2) = V_2 + I$. Since $(x-1)^2 \in V_2$, $x^2 - 1 \in I$, we have

$$(x^2 - 1) - (x - 1)^2 = 2(x - 1) \in \mathrm{Id}(\mathcal{V}_2 \cap \omega\mathfrak{B}_2).$$

On the other hand, take the canonical unitriangular representation

$$\mathrm{ut}_2(\mathbb{F}_2) = (\mathbb{F}_2 \oplus \mathbb{F}_2, \ \mathrm{UT}_2(\mathbb{F}_2))$$

over $\mathbb{F}_2 = \{0, 1\}$ regarded *as a representation over* $\mathbb{Z}$. Clearly this representation is contained in $\mathcal{S}^2 \cap \omega\mathfrak{B}_2 \subseteq \mathcal{V}_2 \cap \omega\mathfrak{B}_2$, and since it is nontrivial, $x - 1 \notin \mathrm{Id}(\mathcal{V}_2 \cap \omega\mathfrak{B}_2)$. It follows that there is 2-torsion in $\mathbb{Z}F/\mathrm{Id}(\mathcal{V}_2 \cap \omega\mathfrak{B}_2)$, whence $\mathcal{V}_2 \cap \omega\mathfrak{B}_2$ is not pure. $\square$

Our concluding remark is concerned with the map $\nu' : \mathrm{M}(K) \to \mathrm{M}(R)$ for an integral domain R and its field of fractions K. Let R be a Dedekind domain. It was proved in the preceding section that in this case ν' is a homomorphism of semigroups. *Is ν' a homomorphism of lattices?* Corollary 4.4.3 combined with Lemma 4.3.2 shows that the answer is negative even over $\mathbb{Z}$.

4.4.4. Corollary. *The map* $\nu' : \mathrm{M}(\mathbb{Q}) \to \mathrm{M}(\mathbb{Z})$ *does not preserve intersection and therefore is not a homomorphism of lattices.* $\square$

4.5. Some applications: an overview

Throughout these notes we have systematically demonstrated that various algebraic theories can be applied to the study of varieties of group representations. The purpose of this final section is to present examples of feedback. We will try to show that the theory of varieties of group representations has interesting applications in other areas of algebra, such as varieties of groups, polynomial identities and varieties of rings, and dimension subgroups. It is an extensive material, so that our exposition is inevitably superficial: most of the proofs are not included and some important results are not mentioned. Nevertheless we hope that this brief overview will illustrate some possibilities of the theory.

ALGEBRAS WITH NILPOTENT COMMUTATOR IDEAL. Consider the variety of (associative K-) algebras defined by the identity

$$(x_1 x_2 - x_2 x_1) \ldots (x_{2n-1} x_{2n} - x_{2n} x_{2n-1}). \tag{1}$$

Is such a variety Specht? In other words, does every algebra whose commutator ideal is nilpotent have a finite basis of identities? This question was
investigated in a number of papers devoted to varieties of rings (see e.g. [20,
53]). It was solved in the affirmative over a field of characteristic zero and
over any infinite field if $n = 2, 3$. Now we can easily show that the answer
is positive in a more general situation.

4.5.1. Corollary (Krasil'nikov [47]). *Every associative algebra over
an infinite field, satisfying the identity (1) for some n, has a finite basis of
identities.*

P r o o f is an immediate consequence of Theorems 4.2.6 and 1.3.1. Let
$\mathcal{M}$ be the variety of algebras over an infinite field determined by (1). The
variety of group representations $\mathcal{M}^\alpha$ defined as in § 1.3 satisfies the identity
(7) from § 4.2. By Theorem 4.2.6 $\mathcal{M}^\alpha$ is Specht and therefore satisfies the
descending chain condition on subvarieties. By Theorem 1.3.1 the map α
is injective, and since it preserves inclusions, $\mathcal{M}$ must satisfy the d.c.c. as
well. $\square$

IDENTITIES OF BLOCK-TRIANGULAR MATRICES. For some time, it has
been unknown whether a finitely generated PI-ring satisfies all identities of
some full matrix $M_n(\mathbb{Z})$ (see e.g. [14', Problem 2.143]). The Razmyslov–
Kemer–Braun theorem on the nilpotency of radical made it possible to
reduce this problem to a question on identities of block-triangular matrices.
To formulate it, let K be an integral domain and let $B_{m,n}(K)$ be the K-
algebra of all (lower) triangular $m \times m$ matrices whose entries belong to
$M_n(K)$. As usual, for any K-algebra A let $T(A)$ denote the T-ideal of its
identities in $K\langle X \rangle$. *Is it true that*

$$T(B_{m,n}(K)) = [T(M_n(K))]^m \ ? \tag{2}$$

For K a field this equality was proved long ago by Lewin [55']. Now
there are known several approaches allowing one to prove (2) over an arbitrary domain. Following [97], we outline one of these approaches. It is
based on the technique of triangular products developed in the theory of
varieties of group representations.

Triangular products were introduced by Plotkin (see e.g. [79]) and can be defined as follows. Let $\rho = (V, G)$ and $\sigma = (W, H)$ be two representations. Then $\Phi = \mathrm{Hom}_K(W, V)$ has a natural structure of H-G-bimodule which allows to define the group of formal matrices

$$N = \begin{pmatrix} G & 0 \\ \Phi & H \end{pmatrix}. \tag{3}$$

This group acts on $V \oplus W$ by the rule

$$(v, w) \circ \begin{pmatrix} g & 0 \\ \phi & h \end{pmatrix} = (v \circ g + w\phi, \ w \circ h). \tag{4}$$

As a result, we obtain a representation $(V \oplus W, N)$ called the *triangular product of ρ and σ* and denoted by $\rho \bigtriangledown \sigma$.

There is a number of results describing the identities of $\rho \bigtriangledown \sigma$ in terms of the identities of ρ and σ. We will need one of them.

4.5.2. Theorem (Vovsi [100]). *Let $\rho = (V, G)$ and $\sigma = (W, H)$ be representations of groups over an integral domain K. If the K-modules V, W, $KF/\mathrm{Id}(\sigma)$ are projective, then*

$$\mathrm{Id}(\rho \bigtriangledown \sigma) = \mathrm{Id}(\sigma)\, \mathrm{Id}(\rho).$$

It is easy to see that the construction of triangular product is quite universal and can be literally carried over to linear representations of other algebraic structures: associative algebras, Lie algebras, semigroups, etc. For example, if $\rho = (V, G)$ and $\sigma = (W, H)$ are representations of (associative K-) algebras, then the triangular product $\rho \bigtriangledown \sigma = (V \oplus W, N)$ defined by (3) and (4) will be also a representation of an algebra. Moreover, all main results dealing with identities of triangular products of group representations remain valid for representations of algebras. In particular, there is an analogue of Theorem 4.5.2 which can be stated as follows.

4.5.2'. Theorem (Vovsi [97]). *Let $\rho = (V, G)$ and $\sigma = (W, H)$ be faithful representations of algebras over an integral domain K. If the K-modules V, W, $K\langle X \rangle$ are projective, then*

$$T \begin{pmatrix} G & 0 \\ \Phi & H \end{pmatrix} = T(H)\, T(G).$$

Now let $\mu_n = (K^n, M_n(K))$ be the natural representation of the full matrix algebra of degree n. Note that it is faithful, K^n is a free K-module and $\Phi = \text{Hom}_K(K^n, K^n) = M_n(K)$. Furthermore, it is well known that the algebra $K\langle X\rangle/T(M_n(K))$ (the so-called algebra of generic matrices) is a free K-module. Hence, by the preceding theorem,

$$T(B_{2,n}(K)) = T\begin{pmatrix} M_n(K) & 0 \\ M_n(K) & M_n(K) \end{pmatrix} = [T(M_n(K))]^2.$$

By induction (see [97] for the details) we obtain the desired result.

4.5.3. Corollary. *If K is an integral domain, then $T(B_{m,n}(K)) = [T(M_n(K))]^m$.*

Using well known arguments (see [55'] and [56']), we can now answer our initial question. Recall that a *Jacobson domain* is an infinite integral domain in which every prime ideal is an intersection of maximal ideals. Let K be a noetherian Jacobson domain and A a finitely generated PI-algebra over K. If J is the Jacobson radical of A, then A/J is a subdirect product of simple algebras Q_i which are finite-dimensional over their centers C_i, and the dimensions of Q_i over C_i are uniformly bounded. Since the K-algebra $M_n(C_i)$ is contained in the variety $\text{var}(M_n(K))$, this implies that $A/J \in \text{var}(M_n(K))$ for some n, that is A/J satisfies all identities from $T = T(M_n(K))$. By the Razmyslov–Kemer–Braun theorem J is nilpotent of some index m, so A satisfies all identities from T^m. By Corollary 4.5.3, $T^m = T(B_{m,n}(K)) \supseteq T(M_{mn}(K))$. Thus:

4.5.4. Corollary. *Every finitely generated PI-algebra over a noetherian Jacobson domain K satisfies all identities of the algebra $M_r(K)$ for some r.*

SOLUBLE JUST-NON-CROSS VARIETIES OF GROUPS. A variety of groups is called *just-non-Cross* if it is non-Cross but every its proper subvariety is Cross. Using Zorn's Lemma, one can easily show that every non-Cross variety contains a just-non-Cross subvariety. Therefore, after the structure of Cross varieties of groups had been characterized in [69] and [44], the problem of investigation of just-non-Cross varieties became one of the most natural.

By the mid-sixties, the following examples of just-non-Cross varieties were known:

(i) the variety $\mathfrak{A}$ of all abelian groups;

(ii) the product $\mathfrak{A}_p\mathfrak{A}_p$ where $\mathfrak{A}_p$ is the variety of abelian groups of exponent p;

(iii) $\mathfrak{A}_p\mathfrak{A}_q\mathfrak{A}_r$ where p, q, r are pairwise distinct primes;

(iv) $\mathfrak{A}_p\mathfrak{Q}_q$ where p, q are distinct primes, and $\mathfrak{Q}_q$ is the variety of 2-nilpotent groups of exponent q if q is odd, and of exponent 4 if $q = 2$.

All these varieties are soluble, and it was conjectured by Kovács and Newman [46'] that the above list contains all soluble just-non-Cross varieties of groups.

In 1971 this conjecture was proved by Ol'shansky [71]. It is a deep and powerful result, and it certainly deserves a separate chapter to be properly presented. We will only give a rough idea of the structure of its proof, trying to emphasise the natural connections with varieties of group representations.

From now on $\mathfrak{V}$ is a soluble just-non-Cross variety of groups. One has to prove that it coincides with one of the above listed. Suppose that $\exp \mathfrak{V} = 0$. Since every variety of exponent 0 contains $\mathfrak{A}$, it follows that $\mathfrak{V} = \mathfrak{A}$. So we may assume that $\mathfrak{V}$ has a nonzero exponent. In view of the solubility, $\mathfrak{V}$ is locally finite. Further, it is proved that if $\mathfrak{V}$ contains nilpotent groups of an arbitrarily large class of nilpotence, then $\mathfrak{V} = \mathfrak{A}_p^2$ for some p.

In the sequel $\mathfrak{V}$ is different from $\mathfrak{A}$ and $\mathfrak{A}_p^2$. In view of the above, $\mathfrak{V}$ is locally finite and the classes of its nilpotent groups are uniformly bounded. A skillful argument shows that *under these conditions* $\mathfrak{V} \subseteq \mathfrak{A}_p\Theta$, *where* Θ *is a Cross variety of groups such that* $p \nmid \exp \Theta$. This fact is a key to what follows because it allows one to reduce the problem to some problem on varieties of group representations, or more precisely, to bivarieties.

A class of group representations $\mathcal{X}$ is called a *bivariety* if it can be defined by identities of two sorts: identities of action

$$y \circ u(x_1, \ldots, x_n) \equiv 0 \qquad (u_i \in KF) \qquad (5)$$

and identities of the acting group

$$f_i(x_1, \ldots, x_n) \equiv 1 \qquad (f_i \in F) \tag{6}$$

(see [74, Appendix]). It follows from the definition that the study of bivarieties can be reduced to the study of varieties of group representations and
varieties of groups, so that this notion does not have a serious independent
value. Nevertheless, it may be useful and convenient in applications, as will
be seen below.

We already know that to each variety of groups Θ there naturally corresponds the variety of representations $\omega\Theta$. Denote now by $\omega_0\Theta$ the class
of all representations of groups from Θ. Then $\omega_0\Theta \subset \omega\Theta$, and $\omega_0\Theta$ is a
bivariety but not a variety. On the other hand, for an arbitrary bivariety
$\mathcal{X}$, the class $\mathcal{X}'$ of all groups admitting a representation in $\mathcal{X}$ is a variety of
groups. Clearly the inclusion $\mathcal{X} \subseteq \omega_0\Theta$ is equivalent to $\mathcal{X}' \subseteq \Theta$.

We return now to the variety $\mathfrak{V}$. It was already noted that $\mathfrak{V} \subseteq \mathfrak{A}_p\Theta$
where Θ is a soluble Cross variety of groups with $p \nmid \exp\Theta$. The next stage of
the proof consists in establishing a connection between subvarieties of $\mathfrak{A}_p\Theta$
and sub-bivarieties of the bivariety $\omega_0\Theta$ over the prime field $\mathbb{F}_p = \mathbb{Z}/p\mathbb{Z}$. If
$\mathfrak{U}$ is a subvariety of $\mathfrak{A}_p\Theta$, then let $\mathfrak{U}^\alpha$ be a bivariety over $\mathbb{F}_p$ generated by
all representations $\rho = (V, G)$ such that $G \in \mathfrak{U} \cap \Theta$ and $V \rtimes G \in \mathfrak{U}$ (where
$V \rtimes G$ is the semidirect product corresponding to ρ). Then $\mathfrak{U}^\alpha \subseteq \omega_0\Theta$ and,
moreover, the following statement holds: α *is a one-to-one correspondence*
between all subvarieties of $\mathfrak{A}_p\Theta$ *and all sub-bivarieties of* $\omega_0\Theta$ *over* $\mathbb{F}_p$.

Further, there is a natural connection between bivarieties over the field
$\mathbb{F}_p$ and bivarieties over its algebraic closure $\overline{\mathbb{F}}_p$; it is based on maps similar
to the maps ν and ν' from §0.4. Together with the above, this yields a
connection between the subvarieties of $\mathfrak{A}_p\Theta$ and the sub-bivarieties of $\omega_0\Theta$
over $\overline{\mathbb{F}}_p$. A greater part of [71] is devoted to the investigation of these
sub-bivarieties. The following result is of decisive importance.

4.5.5. Theorem (Ol'shansky [71]). *Let* $\mathcal{X}$ *be a bivariety over* $\overline{\mathbb{F}}_p$ *such
that*

(a) $\mathcal{X}' = \Theta$ *is a soluble Cross variety of groups whose exponent is not
divisible by* p;

(b) $\mathcal{X}$ contains infinitely many nonisomorphic finite simple representations;

(c) if $\mathcal{Y}$ is a subvariety of $\mathcal{X}$ such that $\mathcal{Y}' \neq \Theta$, then $\mathcal{Y}$ does not possess
  property (b).

Then either $\Theta \subseteq \mathfrak{A}_q \mathfrak{A}_r$ or $\Theta \subseteq \mathfrak{Q}_q$ where p, q, r are distinct primes.

The proof of this theorem is difficult and impressive. After that, it is
relatively easy to prove that if $\mathfrak{V}$ is a just-non-Cross subvariety of $\mathfrak{A}_p\Theta$,
Θ being as before, then either $\Theta \subseteq \mathfrak{A}_p\mathfrak{A}_q$ or $\Theta \subseteq \mathfrak{Q}_q$. Therefore either
$\mathfrak{V} \subseteq \mathfrak{A}_p\mathfrak{A}_q\mathfrak{A}_r$ or $\mathfrak{V} \subseteq \mathfrak{A}_p\mathfrak{Q}_q$. Since two just-non-Cross varieties can not
properly contain each other, the desired result follows:

4.5.6. Theorem (Ol'shansky [71]). *Every soluble just-non-Cross variety coincides with one of the varieties (i) - (iv).*

FAITHFUL REPRESENTABILITY AND DIMENSION SUBGROUPS. Let $\mathcal{X}$ be
a variety of group representations over a ring K. Denote by $\overrightarrow{\mathcal{X}}$ the class
of all groups G admitting a faithful representation which belongs to $\mathcal{X}$.
Evidently, $\overrightarrow{\mathcal{X}}$ is closed under subgroups and Cartesian products. Moreover,
it is easy to show that $\overrightarrow{\mathcal{X}}$ is closed under filtered products and therefore is a
quasivariety of groups [75]. There arises a general problem: *for a given $\mathcal{X}$,
describe the quasivariety $\overrightarrow{\mathcal{X}}$ in group-theoretic terms.* In full generality it is
hardly accessible; the main difficulty is that nobody knows how to extract
the quasi-identities of $\overrightarrow{\mathcal{X}}$ from the defining identities of $\mathcal{X}$.

Since $\overrightarrow{\mathcal{X}}$ is a quasi-variety, in each group G there is a normal subgroup
H smallest with respect to the property $G/H \in \overrightarrow{\mathcal{X}}$. We denote it by
$D_{\mathcal{X}}(G)$, or by $D_{\mathcal{X}}(K, G)$ if the ring K is not clear from the context. Thus

$$G \in \overrightarrow{\mathcal{X}} \iff D_{\mathcal{X}}(G) = 1$$

and the problem of describing the group classes $\overrightarrow{\mathcal{X}}$ is equivalent to describing
the subgroups $D_{\mathcal{X}}(G)$.

It follows from the definition that $D_{\mathcal{X}}(G)$ is equal to the intersection
of the kernels of all representations of G belonging to $\mathcal{X}$, and so it is natural
to say that $D_{\mathcal{X}}(G)$ is the $\mathcal{X}$-kernel of G. Another definition of $D_{\mathcal{X}}(G)$ can

be given in a purely group ring language. Let $I = \mathrm{Id}\,\mathcal{X}$ and let $I(G) = \{u(g_1,\ldots,g_n)\,|\,u \in I,\ g_i \in G\}$. Then $I(G)$ is an ideal of the group ring KG, and the following formula is straightforward:

$$D_{\mathcal{X}}(G) = G \cap (1 + I(G)).$$

Consider now an important particular case when $\mathcal{X} = \mathcal{S}^n$, the variety of n-stable representations. Then $I = \Delta^n$ and $D_{\mathcal{X}}(G) = G \cap (1 + \Delta^n(G))$ is nothing else than the *n-th dimension subgroup of G over K*, usually denoted by $D_n(G) = D_n(K,G)$. It is a classical object which seems to have been first considered by Magnus. An extensive literature is devoted to the study of the subgroups $D_n(G)$ and their characterization in group-theoretic terms (for the details we refer to [27, 29, 73]). In view of the above, this problem is equivalent to characterizing the quasivarieties of groups $\overrightarrow{\mathcal{S}^n}$, and therefore can be treated by methods of varieties of group representations.

The characterization of the dimension subgroups over fields is well known and belongs to Jennings [37, 37'] and Mal'cev [63]. But the dimension subgroups over an arbitrary ring K are little understood, despite the efforts of numerous researchers. The case $K = \mathbb{Z}$ was especially challenging. For more than thirty years it was unknown whether for any G

$$D_n(\mathbb{Z}, G) = \gamma_n(G)$$

where $\gamma_n(G)$ is the n-th term of the lower central series of G. This equality, known as the Dimension Subgroup Conjecture, was finally disproved by Rips [84'] for $n = 4$ (his counterexample was recently generalized by N. Gupta to any $n \geq 4$).

In terms of the classes $\overrightarrow{\mathcal{S}^n}$, the Dimension Subgroup Conjecture means that $\overrightarrow{\mathcal{S}^n} = \mathfrak{N}_{n-1}$, where $\mathfrak{N}_c$ denotes the class of all c-nilpotent groups.[2] The inclusion $\overrightarrow{\mathcal{S}^n} \subseteq \mathfrak{N}_{n-1}$ is an immediate consequence of Kaloujnine's theorem [38], and so the essence of the problem was whether *every $(n-1)$-nilpotent group admits a faithful n-stable representation*. Now, owing to Rips, we know that it is not the case. However, two interesting and closely related questions (raised by Plotkin in about 1970) are still open.

---

[2] Unless otherwise specified, the ground ring is $\mathbb{Z}$.

4.5.7. Problem. *Does every n-nilpotent admit a faithful f(n)-stable representation?* Equivalently: *does there exist a function f(n) such that $D_{f(n)} = 1$ for every n-nilpotent group G?*

Let ω be the first infinite ordinal. A representation $\rho = (V, G)$ is called ω-stable (or residually stable) if $\cap_{n=1}^{\infty} V \circ \Delta^n(G) = 0$. Denote $D_\omega(G) = \cap_{n=1}^{\infty} D_n(G)$.

4.5.8. Problem. *Does every residually nilpotent group admit a faithful ω-stable representation?* Equivalently: *is it true that $D_\omega(G) = 1$ for every residually nilpotent group G?*

Two different formulations of the problems naturally suggest two different approaches to their solution: an "external" one (based on faithful representability), and an "internal" one (based on the group ring technique). In a sense, these approaches go back to Mal'cev [63] and Jennings [37] respectively. The first approach is directly connected with the subject of this book; to illustrate it, let us show for instance how Problem 4.5.7 can be reduced to finite p-groups. Suppose we already know that for all p all finite n-nilpotent p-groups belong to some class $\overrightarrow{S^{f(n)}}$. Being a quasivariety, this class is residually closed. Since a finitely generated nilpotent group is a residually p-group, it follows that this class contains all finitely generated nilpotent groups. But a quasivariety is a local class, and so $\overrightarrow{S^{f(n)}}$ must contain all n-nilpotent groups, as required. A similar argument reduces Problem 4.5.8 to nilpotent p-groups.

We note a few results whose proofs are based on the idea of faithful representability.

4.5.9. Theorem (Kushkuley [50′]). *Let G be a nilpotent group and A its normal subgroup such that G/A is either torsion-free or finite. If A admits a faithful stable representation, then so does G.*

4.5.10. Corollary. *For G and A as above:*
(i) *if $D_s(A) = 1$ for some s, then $D_t(G) = 1$ for some t;*
(ii) *if $D_\omega(A) = 1$, then $D_\omega(G) = 1$.* $\square$

4.5.11. Theorem (Plotkin [75′]). *If G is a residually nilpotent group of finite (Mal'cev special) rank, then $D_\omega(G) = 1$.*

Further development of these results has been obtained by Hartley (see [29, §2] for the details).

Our concluding remark is concerned with higher transfinite dimension subgroups over a field. The transfinite powers of the augmentation ideal $\Delta = \Delta(G)$ are defined by $\Delta^{\alpha+1} = \Delta^\alpha \Delta$ and $\Delta^\lambda = \bigcap_{\alpha<\lambda} \Delta^\alpha$ for a limit ordinal λ; then the corresponding dimension subgroups are $D_\alpha(G) = G \cap (1 + \Delta^\alpha)$. Since for $\alpha \leq \omega$ the group-theoretic characterization of the $D_\alpha(G)$ over a field is well known, there arises a desire to complete the picture and to obtain such a characterization for any α. Without going into detail, we note that this is equivalent to the characterization of naturally defined classes $\overrightarrow{S^\alpha}$ of "generalized stable groups". The following result solves the problem.

4.5.12. Theorem (Vovsi [92′]). *Let α be an arbitrary ordinal. Denote by $\beta = \beta(\alpha)$ the unique ordinal satisfying the relation $\omega^\beta \leq \alpha < \omega^{\beta+1}$, and by $n = n(\alpha)$ the first positive integer for which $\alpha \leq \omega^\beta n$. Then over any ground field*

$$\overrightarrow{S^\alpha} = \overrightarrow{S^n}(\overrightarrow{S^\omega})^\beta.$$

This theorem gives an explicit description of the dimension subgroup $D_\alpha(G)$ for any α and over any field K (if the word "explicit" can in fact be used in conjunction with transfinite numbers!). To illustrate this claim, let for example $\alpha = \omega^3 5 + \omega^2 13 + 43$ and char $K = 0$. Then $\beta(\alpha) = 3$, $n(\alpha) = 5$ and so $\overrightarrow{S^\alpha} = \overrightarrow{S^5}(\overrightarrow{S^\omega})^3$. In terms of dimension subgroups this means that

$$D_\alpha(G) = D_5(D_\omega(D_\omega(D_\omega(G)))).$$

By Mal'cev–Jennings' theorem, for any $n \leq \omega$ we have $D_n(G) = \sqrt{\gamma_n(G)}$, the isolator of the n-th term of the lower central series. Thus

$$D_\alpha(G) = \sqrt{\gamma_5\left(\sqrt{\gamma_\omega\left(\sqrt{\gamma_\omega\left(\sqrt{\gamma_\omega(G)}\right)}\right)}\right)}.$$

Having created such a frightening formula, the author feels that it is time to come to a stop.

REFERENCES

1. Adjan, S. I., *Infinite irreducible systems of group identities*, Izv. AN SSSR Ser. Mat. **34** (1970), 376–384 (Russian).

2. Anan'in, A. Z. and Kemer, A. R., *Varieties of associative algebras with distributive lattices of subvarieties*, Sibirsk. Mat. Zh. **17** (1976), 723–730 (Russian).

3. Bahturin, Yu. A., *"Identities in Lie Algebras"*, Nauka, Moscow, 1985 (Russian).

4. Bahturin, Yu. A. and Ol'shansky, A. Yu., *Identical relations in finite Lie rings*, Mat. Sb. **96** (1975), 543–559 (Russian).

4'. Bass, H., *"Algebraic K-theory"*, Benjamin, New York, 1968.

5. Baumslag, G., *Wreath products and p-groups*, Proc. Cambridge Phil. Soc. **55** (1959), 224–231.

5'. Bergman, G. M., Personal communication, January 1983.

6. Bergman, G. M. and Lewin, J., *The semigroup of ideals of a fir is (usually) free*, J. London. Math. Soc. **11** (1975), 21–31.

7. Birkhoff, G., *On the structure of abstract algebras*, Proc. Cambridge. Phil. Soc. **31** (1935), 443–454.

8. Bryant, R. M., *On s-critical groups*, Quart. J. Math. Oxford **22** (1971), 91–101.

8'. Bryant, R. M. and Newman, M. F., *Some finitely based varieties of groups*, Proc. London Math. Soc. **28** (1974), 237–252.

9. Bryce, R. A., *Metabelian groups and varieties*, Phil. Trans. Roy. Soc. London, Ser.A **266** (1970), 281–355.

10. Cohen, D. E., *On the laws of a metabelian variety*, J. Algebra **5** (1967), 267–273.

11. Cohn, P. M., *Free ideal rings*, J. Algebra **1** (1964), 47–69.

12. Cohn, P. M., *"Universal Algebra"*, Harper and Row, New York, 1965.

13. Cohn, P. M., *"Free Rings and Their Relations"*, Academic Press, London, 1971.

14. Cossey, J., *Laws in nilpotent-by-finite groups*, Proc. Amer. Math. Soc. **19** (1968), 685–688.

14'. *Dniester Notebook* (unsolved problems in ring and module theory), 3rd ed., Math. Institute, Novosibirsk, 1982 (Russian).

15. Evans, T., *The lattice of semigroup varieties*, Semigroup Forum **2** (1971), 1–43.

16. Formanek, E., *The polynomial identities of matrices*, Contemp. Math. **13** (1982), 41–79.

17. Fox, R. H., *Free differential calculus. I*, Ann. of Math. **57** (1953), 547–560.

18. Freese, R. S. and McKenzie, R., *"Commutator Theory for Congruence Modular Varieties"*, Cambridge Univ. Press, Cambridge, 1987.

19. Gaschütz, W., *Über die Φ-Untergruppe endlischer Gruppen*, Math. Z. **58** (1953), 160–170.

20. Genov, G. K., *On the Specht property of certain varieties of associative algebras over a field of characteristic zero*, Dokl. Bulg. AN **29** (1976), 939–941 (Russian).

21. Grätzer, G., *"General Lattice Theory"*, Academic Press, New York, 1978.

22. Grinberg, A. S., *Two notes on varieties of pairs*, Latv. Mat. Yezhegodnik **13** (1973), 64–74 (Russian).

23. Grinberg, A. S., *4-stable varieties*, in "Collection of Papers in Algebra", Riga, 1978, pp. 33–39 (Russian).

24. Grinberg, A. S. and Krop, L. E., *Homogeneous verbal ideals*, in "Collection of Papers in Algebra", Riga, 1978, pp. 21–32 (Russian).

25. Gringlaz, L. Ya., *On locally finite-dimensional varieties of pairs*, in "Certain Questions of Group Theory", Latvian State Univ., Riga, 1971, pp. 19–27 (Russian).

26. Gruenberg, K. W., *"Cohomological Topics in Group Theory"*, Lecture Notes in Math. **143**, Springer-Verlag, Berlin, 1970.

27. Gupta, N., *"Free Group Rings"*, Contemp. Math. **66**, AMS, Providence, 1987.

28. Hall, P., *Nilpotent Groups*, Canadian Math. Congress Summer Sem., Univ. Alberta, Edmonton, 1957.

29. Hartley, B., *Topics in the theory of nilpotent groups,* in "Group Theory: essays for Philip Hall", Academic Press, London, 1984, pp. 61–120.

30. Heineken, H., *Endomorphismenringe und Engelsche Elemente,* Arch. Math. **13** (1962), 29–37.

31. Herstein, I., *"Noncommutative Rings",* MAA, Wiley, 1968.

32. Higgins, P. J., *Lie rings satisfying the Engel condition,* Proc. Cambridge Phil. Soc. **50** (1954), 8–15.

33. Higman, G., *Identical relations in finite groups,* in "Conv. Internaz. di Teoria dei Gruppi Finiti, Firenze 1960", Cremonese, Roma, 1960, pp. 93–100.

34. Higman, G., *Ordering by divisibility in abstract algebras,* Proc. London Math. Soc. **2** (1952), 326–336.

35. Higman, G., *On a conjecture of Nagata,* Proc. Cambridge Phil. Soc. **52** (1956), 1–4.

36. Higman, G., *Some remarks on varieties of groups,* Quart. J. Math. Oxford **10** (1959), 165–178.

37. Jennings, S. A., *The structure of the group ring of a p-group over a modular field,* Trans. Amer. Math. Soc. **50** (1941), 175–185.

37'. Jennings, S. A. *The group ring of a class of infinite nilpotent groups,* Canad. J. Math. **7** (1955), 169–187.

38. Kaloujnine, L., *Über gewisse Beziehungen zwischen einer Gruppe und ihren Automorphismen,* in "Berlin Math. Tagung", Berlin, 1953, pp. 164–172.

39. Kappe, W. P., *Die A-norm einer Gruppe,* Illinois J. Math. **5** (1961), 187–197.

40. Kemer, A. R., *The finite basis property for identities of associative algebras,* Algebra i Logika **26** (1987), 597–641 (Russian).

41. Kleiman, Yu. G., *On the basis of identities of the product of group varieties,* Izv. AN SSSR Ser. Mat. **38** (1974), 475–483 (Russian).

42. Kolchin, E. R., *On certain concepts in the theory of algebraic matrix groups,* Ann. of Math. **49** (1948), 774–789.

43. Kostrikin, A. I., *"Around Burnside",* Nauka, Moscow, 1986 (Russian).

44. Kovács, L. G. and Newman, M. F., *Cross varieties of groups,* Proc. Roy. Soc. London Ser.A **292** (1966), 530–536.

45. Kovács, L. G. and Newman, M. F., *On critical groups*, J. Austral. Math. Soc. **6** (1966), 237–250.

46. Kovács, L. G. and Newman, M. F., *Minimal verbal subgroups*, Proc. Cambridge Phil. Soc. **62** (1966), 347–350.

46′. Kovács, L. G. and Newman, M. F., *Just-non-Cross varieties*, in "Proc. Internat. Conf. Theory of Groups, Canberra 1965", Gordon and Breach, New York, 1967, pp. 221–223.

47. Krasil'nikov, A. N., *On the laws of groups, Lie algebras and associative algebras with nilpotent commutator subobject*, in "XX All-Union Algebr. Conf. Abstracts", Novosibirsk, 1989, p. 72 (Russian).

48. Krasil'nikov, A. N. *On the laws of triangulable matrix representations of groups*, Trudy Moscov. Mat. Obshch. **52** (1989), 229–245 (Russian).

49. Krop, L. E., *Verbal ideals of group rings of free abelian groups over a field of characteristic zero*, in "Certain Questions of Group Theory", Latvian State Univ., Riga, 1971, pp. 34–69 (Russian).

50. Kruse, R. L., *Identities satisfied by a finite ring*, J. Algebra **26** (1973), 298–318.

50′. Kushkuley, A. H., *On finitely stable representations of nilpotent groups*, Latv. Mat. Yezhegodnik **16** (1975), 39–45 (Russian).

51. Kushkuley, A. H., *On identities of finite-dimensional representations*, in "Collection of Papers in Algebra", Riga, 1978, 134–157 (Russian).

52. Latyshev, V. N., *On the Specht property of certain varieties of associative algebras*, Algebra i Logika **8** (1969), 660–673 (Russian).

53. Latyshev, V. N., *Partially ordered sets and nonmatrix varieties of associative algebras*, Algebra i Logika **15** (1976), 58–70 (Russian).

54. Levi, F. W., *Groups in which the commutator operation satisfies certain algebraic conditions*, J. Indian Math. Soc. **6** (1942), 87–97.

55. Lewin, J., *Free modules over free algebras and free group algebras: the Schreier technique*, Trans. Amer. Math. Soc. **145** (1969), 455-465.

55′. Lewin, J., *A matrix representation for associative algebras. I*, Trans. Amer. Math. Soc. **188** (1974), 293–308.

56. L'vov, I. V., *On varieties of associative rings. I*, Algebra i Logika **12** (1973), 269–297 (Russian).

56′. L'vov, I. V. *Braun's theorem on the radical of a finitely generated PI-algebra*, Preprint 63, Math. Institute, Novosibirsk, 1984 (Russian).

57. Magnus, W., *Beziehungen zwischen Gruppen und Idealen in einen speziellen Ring*, Math. Ann. **111** (1935), 259–280.

58. Magnus, W., *Über Beziehungen zwischen höheren Kommutatoren*, J. Reine Angew. Math. **177** (1937), 105–115.

59. Magnus, W., *On a theorem of Marshall Hall*, Ann. of Math. **40** (1939), 764–768.

60. Magnus, W., Karras, A. and Solitar, D., *"Combinatorial Group Theory"*, Wiley–Interscience, New York, 1966.

61. Mal'cev, A. I., *On algebras with identical relations*, Mat. Sb. **26** (1950), 19–33 (Russian).

62. Mal'cev, A. I., *On certain classes of infinite soluble groups*, Mat. Sb. **28** (1951), 567–588 (Russian).

63. Mal'cev, A. I., *Generalized nilpotent algebras and their adjoint groups*, Mat. Sb. **25** (1949), 347–366 (Russian).

64. Mal'cev, A. I., *"Algebraic Systems"*, Nauka, Moscow, 1970 (Russian).

65. Medvedev, Yu. A., *The finite basis property of varieties with a two-term identity*, Algebra i Logika **17** (1978), 705–726 (Russian).

66. Nagata, M., *On the nilpotency of nilalgebras*, J. Math. Soc. Japan **4** (1952), 296–301.

67. Neumann, B. H., *Identical relations in groups. I*, Math. Ann. **114** (1937), 506–525.

68. Neumann, H., *"Varieties of Groups"*, Springer-Verlag, Berlin, 1967.

69. Oates, S. and Powell, M. B., *Identities of finite groups*, J. Algebra **1** (1964), 11–39.

70. Ol'shansky, A. Yu., *On the problem of finite basis of identities in groups*, Izv. AN SSSR Ser. Mat. **34** (1970), 376–384 (Russian).

71. Ol'shansky, A. Yu., *The soluble just-non-Cross varieties of groups*, Mat. Sb. **85** (1971), 115–131 (Russian).

72. Ol'shansky, A. Yu., *On the orders of free groups of locally finite varieties*, Izv. AN SSSR Ser. Mat. **37** (1973), 89–94 (Russian).

73. Passi, I. B. S., *"Group Rings and Their Augmentation Ideals"*, Lecture Notes in Math. **715**, Springer-Verlag, Berlin, 1979.

73'. Passman, D. S. *"The Algebraic Structure of Group Rings"*, Wiley, New York, 1977.

74. Plotkin, B. I., *"Groups of Automorphisms of Algebraic Structures"*, Nauka, Moscow, 1966 (Russian).

75. Plotkin, B. I., *Varieties and quasivarieties associated with group representations*, DAN SSSR **196** (1971), 527–530 (Russian).

75'. Plotkin, B. I., *Notes on stable representations of nilpotent groups*, Trudy Moskov. Mat. Obshch. **29** (1973), 191–206 (Russian).

76. Plotkin, B. I., *Varieties of group representations*, Uspekhi Mat. Nauk **32**, No. 5 (1977), 3–68 (Russian).

77. Plotkin, B. I., *Varieties in representations of finite groups. Locally stable representations. Matrix groups and varieties of representations*, Uspekhi Mat. Nauk **34**, No. 4 (1979), 65–95 (Russian).

78. Plotkin, B. I., *Locally finite and locally finite-dimensional varieties of pairs – group representations*, in "Collection of Papers in Algebra", Riga, 1978, pp. 185–245 (Russian).

79. Plotkin, B. I. and Grinberg, A. S., *On the semigroups of varieties associated with group representations*, Sibirsk. Mat. Zh. **13** (1972), 841–858 (Russian).

80. Plotkin, B. I. and Vovsi, S. M., *"Varieties of Group Representations: General Theory, Connections and Applications"*, Zinatne, Riga, 1983 (Russian).

81. Pollák, G., *On the consequences of permutation identities*, Acta Sci. Math. (Szeged) **34** (1973), 323–333.

82. Powell, M. B., *Identical relations in finite soluble groups*, Quart. J. Math. Oxford **15** (1964), 131–148.

83. Processi, C., *"Rings with Polynomial Identities"*, Marcel Dekker, New York, 1973.

84. Razmyslov, Yu. P., *"Identities of Algebras and Their Representations"*, Nauka, Moscow, 1989 (Russian).

84'. Rips, E., *On the fourth integer dimension subgroup*, Israel J. Math. **12** (1972), 342–346.

85. Rowen, L. H., *"Polynomial Identities in Ring Theory"*, Academic Press, New York, 1980.

86. Shevrin, L. N. and Volkov, M. V., *Identities of semigroups*, Izv. VUZ. Mat., 1985, No. 11, 3–47 (Russian).

87. Suprunenko, D. A., *"Matrix Groups"*, Nauka, Moscow, 1972 (Russian).

88. Vaughan-Lee, M. R., *Uncountably many varieties of groups*, Bull. London Math. Soc. **2** (1970), 280–286.

89. Vaughan-Lee, M. R., *Abelian-by-nilpotent varieties*, Quart. J. Math. Oxford **21** (1970), 193–202.

90. Volkov, M. V. and Gein, A. G., *Identities in nilpotent-by-finite Lie rings*, Mat. Sb. **118** (1982), 132–142 (Russian).

91. Vovsi, S. M., *Identities of finite pairs*, in "XII All-Union Algebr. Coll. Abstracts. I", Ural State Univ., Sverdlovsk, 1973, p. 14 (Russian).

92. Vovsi, S. M., *On critical representations of groups*, Latv. Mat. Yezhegodnik **20** (1976), 141–159 (Russian).

92'. Vovsi, S. M., *Prevarieties of generalized stable groups*, Sibirsk. Mat. Zh. **19** (1978), 267–279 (Russian).

93. Vovsi, S. M., *On locally finite varieties of group representations*, Izv. VUZ. Mat., 1979, No. 6, 14–25 (Russian).

94. Vovsi, S. M., *"Triangular Products of Group Representations and Their Applications"*, Progress in Math. **17**, Birkhäuser, Boston, 1981.

95. Vovsi, S. M., *On relatively free group representations of finite rank*, Latv. Mat. Yezhegodnik **28** (1984), 162–164 (Russian).

96. Vovsi, S. M., *The Fox derivatives and varieties of group representations*, Sibirsk. Mat. Zh. **27** (1986), 34–41 (Russian).

97. Vovsi, S. M., *Triangular products and identities of certain associative algebras*, Algebra i Logika **26** (1987), 27–35 (Russian).

98. Vovsi, S. M., *Homogeneous varieties of group representations*, Sibirsk. Mat. Zh. **29** (1988), 35–47 (Russian).

99. Vovsi, S. M., *Two notes on homogeneous varieties of group representations*, Latv. Mat. Yezhegodnik **33** (1989), 84–91 (Russian).

100. Vovsi, S. M., *Varieties and triangular products of projective group representations*, Sibirsk. Mat. Zh. **21** (1980), 56–62 (Russian).

101. Vovsi, S. M. and Nguyen Hung Shon, *Identities of stable-by-finite representations of groups*, Mat. Sb. **132** (1987), 578–591 (Russian).

102. Vovsi, S. M. and Volkov, M. V., *On multilinear identities of group representations*, Commun. Algebra **18** (1990), 389–401.

103. Wehrfritz, B. A. F., *"Infinite Linear Groups"*, Springer-Verlag, New York, 1973.

104. Zariski, O. and Samuel, P. *"Commutative Algebra"*, Van Nostrand, Princeton, 1958.

105. Zel'manov, E. I., *On Engel Lie Algebras*, Sibirsk. Mat. Zh. **29** (1988), 112–117 (Russian).

106. Zhevlakov, K. A., Slin'ko, A. M., Shestakov, I. P. and Shirshov, A. I., *"Rings That are Nearly Associative"*, Nauka, Moscow, 1978 (Russian).

INDEX

A

acting group, 2
action, 1
Adjan, S. I., 113
algebraic closure operation, 156
  with finite basis property, 156
almost surjective homomorphism,
  169
Amitsur–Levitzky Theorem, 9
axiomatic rank, 18

B

basis rank, 19
Bergman, G. M., xiii, xiv, 49,
  147, 165, 166
Birkhoff, G., ix
Birkhoff Theorem, 13
bivariety, 183
Bryant, R. M., 110
Bryce, R. A., x
Burnside variety, 38

C

Cartesian product of representations, 4
category of representations, 3
Cherlin, G., xiv
closed class of simple representations, 108
closure operation, 4
Cohen, D. E., 146, 156, 159, 161
Cohn, P. M., 94
comonolith, 101

completely invariant ideal, 6
Cross, D. C., 100

D

Dedekind domain, 168
dimension subgroup, 186
Dimension Subgroup Conjecture,
  186
direct product of representations, 12

E

endomorphism of representations, 3
Engel identity, 76
equivalence of representations, 3
exponential growth of a function, 97

F

factor (=section) of a representation, 100
  proper, 100
factor-representation, 3
faithful image of a representation, 2
Fox, R. H., xiii, 23
Fox derivation, 24
  of higher degree, 25
Fox fundamental formula, 25
fractional ideal, 167
  invertible, 167
fully invariant ideal, 6

G

Gaschütz, W., 121
generalized euclidean ring, 172

Printed in the United States
By Bookmasters